Survival 101

Raised Bed Gardening

and

Food Storage:

The Complete Survival Guide to Growing Your Food, Food Storage, and Food Preservation in 2021

(2 Books IN 1)

Table of contents

Survival 101: Raised Bed Gardening

The Essential Guide To Growing Your Own Food In 2021 by Rory Anderson

Survival 101: Food Storage

A Step by Step Beginners Guide on Preserving Food and What to Stockpile While Under Quarantine

Survival 101: Raised Bed Gardening 2021

Rory Anderson

Introduction

Raised garden beds are a backyard gardener's secret to happy, healthy plants. When it comes to gardening in a small space, raised bed gardening allows you to have greater control over the quality of your soil. It also allows more space to grow things in, and they are generally easier to tend to. Another great benefit of raised bed gardens is that they can help extend your growing season since they tend to warm up faster than the actual ground. This means you can start crops sooner, run them longer, and get higher yields out of your plants.

Due to the coronavirus, there has never been a better time to get involved with raised bed gardening. While supply chains crash, markets destroyed, and the system falls apart, raised bed gardens can keep you and your family consuming delicious food, regardless of what is going on in the world. The key is to learn quickly and start as soon as possible so that you have enough food stored to keep you and your family going through the winter of 2021.

While *Survival 101: Raised Bed Gardening 2021* is handy for a hobby gardener, you should know that the real reason this book was written was to aid you in your survival. If you need to rely on your garden for self-sufficiency, you will discover everything to do that here. This way, no matter what happens now or in your future, you can feel confident that you will be able to feed yourself and your family through your garden.

If you are brand new to gardening, I encourage you to pay close attention. The content of this book is perfect for anyone with any skill level. Everything is clearly defined so that you know exactly what to do, when, and how. This way, there is no guessing game to be played, and you can grow your garden successfully.

In some areas of this book, you will notice that certain aspects may not be entirely relevant to your climate or geographical region. I encourage you to read through them, anyway. Educating yourself on how to garden anywhere ensures that no matter what happens in your life, you have the knowledge you need to thrive.

If you are ready to discover the art and power of raised bed gardening, let's go!

Chapter 1

Raised Bed Gardening Basics

Before you do anything, you must know the basics of what it takes to make a raised bed garden work. Proper planning is essential in any survival or subsistence endeavor. With raised bed gardens, having the proper location, elements, and logistics will ensure that your garden will thrive.

As with most things, proper preparation ensures success because it prevents you from making terrible mistakes that would be far more challenging to rectify down the line. Once your garden is built, the soil is set, and plants are growing, you will not be able to easily move things around and make adjustments without risking the loss of your plants.

The four basic things to consider when it comes to raised bed gardening include sun, drainage, water, and logistics.

Sun

The majority of plants require sunlight as an essential aspect of their growth. Sunlight allows plants to make their food through the process of photosynthesis. You have likely heard of this process before, but in case you haven't, I will quickly explain it to you. In rich growing soil, there are a variety of different nutrients that are essential to a plant's growth. When the soil becomes wet, the water absorbs the nutrients. The plants then absorb that water through their roots, which introduces the nutrients into their system. However, they need energy to use those nutrients for growth. They get that energy from the sunlight. The green pigment in the leaves of plants, called chlorophyll, traps the sun's energy and allows it to fuel the process of breaking down the food absorbed through the soil, effectively "feeding" the plant. This is the plant's version of a digestion system!

Most herbs, vegetables, and fruit plants will require at least 6 hours of sunlight per day. This sunlight needs to be direct, which means it needs to land directly on their leaves. Bright indirect light or filtered light will not be sufficient to grow your plants.

Raised bed gardens are excellent for helping plants gain better access to sunlight. In many urban environments, even a south-facing garden can have shadows covering large portions of the yard. Raising your garden into raised garden beds means that your plants lie above the shadows, creating direct access to the sun's rays.

Drainage

Water is an essential part of the plant's digestive system, yet too much water is not a good thing. Soil with low-quality drainage can lead to water becoming trapped in the soil underneath the plants. This can lead to issues with excess water build up which can cause a variety of problems that will kill your plants. One problem is root rot, which is caused by the roots never having a chance to dry out between feedings. Another problem that can be caused by poor drainage is essentially a plant's version of drowning. That is, the plant absorbs far too much liquid and the cells within the plant burst and die, causing the plant itself to die.

Proper drainage is hard to facilitate in the ground where you cannot reasonably control what is going on beneath your garden bed. While you can turn the topsoil, it would be entirely unreasonable to dig down 2-3 feet into your garden to add something for improved drainage. Not only would this be a large amount of work, but it may not even lead to the results you desire.

Raised bed gardens are designed above the ground, which means you have complete control over the drainage of your garden. You can lie rocks or bricks across the bottom of the bed before adding soil as a way to provide extra space for drainage. This way, even if you live in an area where soil drainage is naturally poor, you can still have excellent drainage that will allow your plants to thrive.

Water

Since water plays such a vital role in a plant's ability to absorb nutrients, you will need to keep a steady supply of it coming into your raised bed garden. With raised bed gardens, the soil tends to dry out quicker because there is a smaller volume of it, which means moisture evaporates faster. The watering aspect of your raised bed garden requires you to ensure that your garden is getting enough water for it to thrive.

Typically, vegetable gardens should be watered at least once a day for approximately ten to fifteen minutes. This provides the plants with plenty of water to grow and bear strong, healthy vegetables. You should always water your garden in the evening after the shade has covered it, and the afternoon heat has dissipated. Watering your garden in the morning or the afternoon can result in your vegetable plants burning, as the excess water on the leaves and in the soil will raise the temperature too high.

It is best to ensure that you have a reasonably sized hose that you can use to reach all, or at least most, of your garden. This way, you will not have to carry multiple watering cans back and forth across your yard to water all of your plants. Use the showerhead setting on your hose to ensure that the water is evenly spread across the soil like rainfall, rather than having single small areas being drenched by a heavy flow. If your hose does not reach across your garden, consider using a wagon and having 3-4 watering cans. Then you can fill all of them and wheel them over to the less accessible parts of your garden.

Logistics

The last thing you have to consider is the logistics of your garden. There are different logistics to be considered; where your garden is, how large your garden is, and whether or not your garden obeys your municipality's bylaws.

Having your garden appropriately placed is vital to ensuring that it gets adequate access to sunlight and that you can easily bring water to the garden. In addition to having the garden itself placed correctly, you need to make sure that the plants inside of your garden are placed properly, too. Ideally, your garden should be in a south-facing lot. Your tallest plants should be planted on the north side of your garden beds, and the shorter plants should be planted on the east side. This way, your garden gets plenty of light, and your taller plants do not crowd or cast shadows over your smaller plants as they grow.

The size of your garden will depend on what you have available and what you can afford. While we will go into more detail on specific sizing and development considerations in Chapter 2, it is worth noting that this is a logistics that need to be planned *before* you start building your raised bed garden. The size of your garden will affect how easy it is to care for, how much food you will be able to grow, and how effective your garden will be at sustaining your family's needs.

Lastly, you need to consider your local bylaws. Some municipalities have bylaws or regulations preventing you from having raised bed gardens in certain parts of your yard. They may also regulate how you can water your garden, and what you can water your garden with. Having clarity about your local bylaws ahead of time will ensure that you are within regulation and protect you from being asked to take your garden apart at any point during the growing season.

Chapter 2

The Size of Your Garden

The right sized garden is one that will give you a significant yield without being too challenging for you to manage. Finding the perfect size for your garden is essential in ensuring that you have a thriving crop through the season. Understand that what works for one person might not work for everyone. You have to honestly consider what you are capable of, what you have access to, and what would be reasonable for you to manage.

How Big Your Garden Beds Should Be

While there is quite a bit of flexibility with raised bed gardening, there are some specifics that need to be met. Raised bed gardens should be at least 3ft by 3ft; otherwise, you will not have enough space for your garden to yield anything significant. If you would prefer to have larger garden beds, you can choose to make them as long as you want; however, they should be no wider than 4ft. If your garden gets wider than 4 ft, it will become challenging to manage. The idea is that you can easily access either side of the garden this way. This will allow you to freely water everything, minimize weeds and pests, and properly take care of your plants. If your garden becomes too large, you will struggle to maintain it, and the plants themselves will struggle to thrive, too.

Aside from minimum size considerations, you are going to need to determine what amount of garden space is going to be optimal for your family. The size should be based on how many crops you want to grow and how many plants from each crop you want to grow. Once you know this information, you can identify the space requirements for each plant and perform your calculations accordingly. Through this, you will discover how much space you need. If you come up with a number that seems unreasonable or unachievable based on the amount of space you have, you should adjust your crop plans. Do this to accommodate for what you can reasonably accomplish in your present situation.

The exact width of your bed should also be determined based on who is going to managing your garden. If the person tending to your garden is shorter or has shorter arms, you will want to stick to the 3ft width to ensure easy access. If the person tending to your garden is taller or has a more effortless ability to move around, expanding your garden to a 4ft width will allow for even more crops to grow while still being manageable for the person tending it.

The depth of your raised garden beds needs to be determined based on the types of crops you are growing in them. The absolute minimum soil depth in any raised bed garden is 6 inches. Certain crops, such as carrots and potatoes, will require deeper beds because they grow deeper into the soil. Having your gardens adequately raised to accommodate for this ensures that your deeper crops will have enough space to flourish.

You also need to consider the amount of space you leave between your raised bed gardens. Generally, raised bed gardens are built in rows so that plenty can be developed, and yet there is still space for you to walk between the gardens and tend to your crops. When you are planning space for your walkways, consider anyone who may need access to these walkways, as well as any tools you may need to bring down them. For example, if you will be using a wagon to transport water or supplies around your garden, you need walkways large enough to move that wagon around.

As you begin to develop plans for your garden, do not feel like you need to create everything in a perfect square or rectangle. You can build your gardens in any shape you desire, based on the space you have available to you. This means if you need an unusual garden so that your garden adequately fits your space, you can make a unique garden. As long as it is no longer than 4ft wide and deeper than 6 inches, you can prepare whatever you need for your raised bed gardens.

How Big Your Crop Should Be

There are many things to take into consideration when it comes to determining what crops you grow, and how large your crops should be. Aside from the obvious factor of space, you need to consider what your family needs, what you can reasonably achieve, and what is going to give you the best range of diversity.

Chances are, your urban garden is not going to provide your family with enough fruits and vegetables to last you through an entire year. Unless you have a particularly large space, attempting to replace your need to rely on external sources for fruits and vegetables altogether is not feasible. Instead, you want to focus on substituting part of your diet through the support of your garden. This way, you can rely *less* on grocery stores and other food sources, and rely more on yourself. Do not mistake this for meaning that you cannot get a good, long-lasting yield out of your raised bed gardens, though. Raised bed gardens tend to yield higher crops than in-ground gardens. You can produce enough that you can eat it through the growing season and preserve some to eat at a later date, too.

Diversity matters when it comes to growing a garden, too. Growing too much of any one thing will leave you with plenty of that produce on hand, but nothing else. This can be boring for your palate, as well as poor for your health as you require diversity in your produce to obtain maximum nutrients through your diet. If you do not grow a diversified crop, you will still have to rely heavily on external sources for your food supply. Growing your diversified crop ensures that you have a healthy range of things to eat. As well as for the growing season, you will not have to rely on the grocery store nearly as much.

When it comes to determining what crops to grow, you'll need to think about what you are most likely to eat, what will be easiest for you to grow in your space and with your skill level, and what will be easiest for you to preserve out of your garden. The crops that you will eat, that you can make thrive, and that you can preserve for long term consumption are the best crops for you to pick for your raised bed gardens. From there, pick as much as you can reasonably grow, or as much as you can honestly eat in a year, whichever is less.

Managing Crop Sizes As a Beginner

If you are brand new to gardening, it is not ideal for planning to create the largest garden possible with the most crops possible. While it is a noble idea, this plan can rapidly lead to overwhelm and crop loss. Although gardening itself may seem simple, there is a learning curve that comes with the practice, and that learning curve can take some time to get the hang of. Trying to grow too much variety at once or too many plants can lead to you struggling to identify the proper care methods for each plant. This can result in each of the plants failing to thrive. What ends up happening, then, is you do not see a significant yield of anything.

Rather than overwhelming yourself, start small. Begin by picking just a few different fruits or vegetables that you enjoy, and that will grow in your growing zone and plant those. Learn how to take proper care of them throughout the growing season. As well, learn how to properly harvest, cook, and preserve them. The following year, when you feel more comfortable with the original plants, you can grow them again and add a few new varieties. Continue growing out until you find yourself comfortably managing larger crops with greater diversity to them.

Chapter 3

Choosing and Planning Your Plants

Picking the right plants for your raised bed gardens ensures that you have the best chance at a thriving garden, as well as a vast range of foods to consume through the growing season and beyond. Before you lay any seeds in your garden, make sure you pick the right plants and that you determine the correct number of plants per crop. This ensures that you get the best range of diversity and yield out of your garden.

In addition to choosing what plants you are going to grow, you need to have a clear plan as to when and how much. Gardeners and farmers alike often keep what is known as a "growing chart," which allows them to keep everything organized. This chart will tell them what to plant, when, and how much. It will also tell them which raised garden bed they should plant their crops in so that they know exactly where it belongs. Keeping a clear, organized outline for how your growing season plan ensures that you can keep yourself on track for an excellent harvest every single year.

Before you can get into creating your growing chart, you need to know a few things about planning a garden. This way, your chart is designed with functionality in mind, and it is truly fit for helping you cultivate a healthy garden with a high crop yield.

Keeping a Plant Journal

Plant journals are a great way to help you stay organized with what you are trying, what is working, and what you want to try next. Your journal is different from your chart because it is focused less on planning and more on tracking your progress and anything you learn along the way. Keeping a plant journal ensures that you never make the same mistake twice. It also helps you remember which new techniques you wanted to try in each succeeding growing season.

Over time, your plant journal will help you master your favorite crops and grow your garden with great success.

Inside of your plant journal, keep a document of every single plant you grow in a season. Be sure to write down specific details about that plant, including what it is, where you placed it, and how you cared for it, too. Keep track of each time the plant is watered, fertilized, pruned, and harvested so that you know how well your care efforts are working. As well, keep track of how much you harvested. With plants, many things will work, but often only a few things will truly increase your yield. Since you want to have the biggest yield possible, you want to identify those exact techniques that get you the biggest yield, and then you want to practice them year after year. This way, you are always getting as much as you can out of your plants.

Another section you might wish to keep in your plant journal would be dedicated to plants you want to try growing, or techniques you want to try using. These days, researching plants and plant care is easily accessible. You are likely going to stumble across many different things regarding different plants you want to try or techniques you want to try. Keeping track of those things in your journal will allow you to review them, test them, and see if they work for you. If they don't, keep track of that to avoid making the same mistake twice.

Companion Planting

Companion planting is a practice where farmers plant different types of plants together because they complement each other and promote better growth within one another. This is a technique that has been used since the pioneer times because it provides gardeners with many benefits. First and foremost, this maximizes your yield while minimizing the amount of space that is required to grow your plants. Many times, companion planting can lead to organic gardening benefits such as natural pest control or natural fertilizer due to the types of nutrients that each plant contributes to the soil.

Ensure that you continue to respect each plant's unique needs while growing companion plants together. For example, you still need to have taller plants on the north side of the bed, and the plants need to be spaced out enough to support healthy root growth. While these plants will benefit each other, they still have individual requirements that need to be met for them to thrive.

To help you get started, I have included a list of companion plants you can grow together. This list includes many common vegetables, including ones that you will likely grow in your garden.

- Beans, Corn, and Squash: These are called the "three sisters" and are the oldest companion planting technique out there. Corn provides beans with something to grow up. Beans provide nitrogen to the soil, while corn uses a lot of nitrogen to grow. Squash has large leaves that block weeds from growing on the soil. Because of all of this, the three grow wonderfully together.
- Beets: Beets can grow with broccoli, bush beans, cabbage, garlic, kale, lettuce, or onions. They cannot grow with pole beans, though, as they will stunt each other's growth.
- Cucumbers: Cucumbers can grow with bush beans, cantaloupe, lettuce, and radishes. They cannot grow with potatoes, though, as cucumbers tend to cause potato blight.
- Lettuce: Lettuce grows well with carrots, but not cabbage. While lettuce has shallow roots, carrots have deep roots, so they do not compete with each other for space.

- Onions: Onions can be grown with asparagus, broccoli, Brussels sprouts, cabbage, and kale. They cannot be grown with beans or peas. Onions are known to naturally repel pests such as cabbage worms, making these two excellent growing companions.
- Peas: Peas grow well with carrots, radishes, and turnips. They do not grow well with beans, garlic, or onions. With peas and beans, they are not compatible on a chemical level, and growing them together can lead to them tasting strange.
- Peppers: Peppers can be grown with basil, carrots, eggplants, onions, parsley, and tomatoes. They cannot be grown with kohlrabi. Basil, in particular, keeps flies and mosquitoes at bay, both of which can destroy pepper plants.
- Potatoes: Beans, corn, and peas grow well with potatoes. Peppers and tomatoes do not. With potatoes, tomatoes, and peppers, they are all part of a family of vegetables known as "nightshades." This means they are all prone to the same diseases. Having them too close together means that if one plant gets a disease, it will rapidly spread through all three crops and wipe them out.
- Tomatoes: Tomatoes plant well with asparagus, basil, and marigolds. They do not work well with corn. Asparagus deters nematodes, basil repels flies and mosquitoes, and marigolds keep away hornworms, which can all support you in protecting your crop.

Growing Seasons

Each plant grows best during a specific time of the year. The time of year contributes to proper temperatures, proper precipitation and humidity levels, and proper light levels. Plants will rarely, if ever, grow outside of their growing season unless you have a greenhouse to stimulate specific growing conditions.

As you plan your crops for the year, be sure to consider which part of the growing season each plant grows best in and plant your crops accordingly. This way, your plants get the growing conditions they need without you needing to create those growing conditions artificially. As well, you can maximize your crop yield from each plant, giving you the best results possible. Understand that just because you plant your plants in their respective growing season does not mean they will be ready to harvest in that same season. Garlic, for example, is a fall crop because it is planted in the fall, but you will not harvest garlic until the spring.

The growing season is generally broken down by spring and fall. The harvest you gain throughout the summer is from the plants that you grew in the spring. Generally, no new plants are added to a garden after early June.

Below, I have included a list of some of the vegetables you can grow in each growing season to help you get an idea of what to plant and when. However, this chart is not specific to every single zone. You are going to need to consider your growing zone, as well, to get exact timeframes on what to plant and when.

- Spring: Beans, beets, broccoli, Brussels sprouts, cabbage, carrots, cauliflower, corn, cucumber, lettuce, onions, peas, peppers, spinach, squash, tomatoes.
- Fall: Beans, beets, broccoli, cabbage, carrots, cauliflower, corn, cucumber, kale, lettuce, peas, peppers, spinach, tomatoes.

If you are unsure about what the growing season looks like for your area, look for resources from the USDA. The USDA offers free gardening education, which will include information on how to plant for your zone, and when.

Your Local Growing Zone

As you know, each plant needs a specific amount of light, water, humidity, heat, and other essentials to help it grow. Each part of the world has its own unique climate, contributing to its ability to provide specific light, water, humidity, and temperatures. Not all geographical regions are capable of offering the same growing conditions; therefore, not all plants thrive in all areas of the world. Growing zones are designed to inform you which plants will work best in your area.

The USDA offers maps that help you identify which zone you are in and which plants will grow in your zone. You must check your growing zone for planting information to ensure that you are growing crops that will thrive in your garden. Attempting to grow plants that are not known for growing in your zone can result in the plant failing to grow, or the crops being incredibly poor. There is no point putting effort into growing plants that are unlikely to thrive, so opt instead for plants that are going to grow well in your zone.

If you want to take your gardening abilities to the next level, do not just look for a list of what you can grow in your area. Research what your growing zone is like so that you clearly understand what specific conditions your zone offers. This will help you understand how you need to supplement your zone's natural conditions to ensure that your plants thrive.

Plant Spacing

In-ground gardens often have much denser soil conditions, which means that it is tougher for vegetable roots to grow down into the soil. Raised bed gardens, however, are loosely packed, which means that the roots can easily grow down rather than out to the sides. This means that you can plant more plants in a raised bed garden than you can in an in-ground garden because the surface area is less important than the depth. In a conventional garden, plants must be kept at least 3 feet apart. In a raised bed garden, you can get away with 1 foot of space between most plants, or less depending on what you are growing. Carrots, for example, can be spaced just 2-3 inches apart in their own garden plot. Larger plants, such as potatoes, will need about 15-18 inches of space, which is still much closer than the 3 feet they would require in a conventional in-ground garden.

Aside from root spacing, you also need to consider spacing for the above-ground portion of the plant. For example, squash plants may require minimal root space, but the plant grows like a vine, which requires an abundance of space. Some vine plants can be grown up garden trellises, allowing you to continue maximizing your space. Others, like pumpkin plants, may be too heavy for that. For that reason, be sure to provide them with an abundance of crawling space to ensure that they thrive without becoming crowded or crowding anything else in your garden.

Plant Arrangement

Your plants should always be arranged based on what size they will grow to be when they reach above ground, too. Taller and bushier plants can cast shadows on plants behind them, so keeping them along the north side of your garden is important. Your garden should get increasingly shorter until you reach your shortest plants on the south side of your garden.

Aside from height, also consider how you are going to organize your plants into your raised bed gardens. There are two primary layouts you can use to help you decide where plants are going to go in your garden. One layout includes you planting things in rows, while the other includes you sectioning off portions of your garden beds for certain plants. For example, you might use the left half of a raised bed for carrots and lettuce, while you might use the right half for onions and kale. Find what works best for you, and what is going to make the most sense for your space, then go from there.

For vine plants like beans and squash, make sure you plan an area for them to grow in advance. Having adequate space for them to grow into ensures that you can maximize your yield from these plants. If you are unable to offer ground space for squash, cucumbers, and other crawling plants, consider using a heavy-duty trellis, loose mesh bags, and hooks. You can encourage the vine of these plants to crawl up the trellis. Then, anytime you see fruit growing from them, you can put it in a fabric mesh bag and hook it to the trellis. This way, the fruit does not break off the vine before it has a chance to grow to maturity. Be sure to check on fruits in the bags to avoid having them go rotten inside of the bags. Using loose mesh is best as it will allow plenty of breathing space and room for growth.

Your Growing Chart

After you have completed all of this research work, you can develop your growing chart. The easiest way to develop your growing chart is to take a notebook and write it out one month at a time. For each month, write down what you are going to plant, where you are going to plant it, and how you are going to tend to it. Also, keep a note of when each plant will harvest and plan that into your monthly growing chart so that you are aware of when you need to be ready to collect, cook, and preserve those items. For example, if you plant pole beans in March, you know that the first harvest should be ready in May. In your March growing chart, you would write down your intention to plant the pole beans, while in your May growing chart, you would write down your intention to harvest the pole beans.

Complete your growing chart at least one to two months before the growing season starts so that you are ready for it. This gives you plenty of time to collect your seeds, tend to your soil, and prepare your essentials, so that come growing season, you are ready to go. Once the growing season starts, you do not want to have to worry about planning last-minute crops or dealing with last-minute preparation work. Every minute you spend changing the plan is a minute where you take away from the harvest you are already working toward. Have a clear plan and see it all the way through. The only time your plan should change is if it is imperative that it change, in which case you should stop and complete a new plan for the problem you have been presented with. Then, once you have a clear plan to follow, you should follow that just as loyally as your original plan. This is the best way to ensure that you never accidentally destroy your harvest due to an impulse decision that might have larger consequences than you anticipated.

Chapter 4

Crop Rotation

To understand the purpose of crop rotation, you first need to understand what happens to soil after plants have been "feeding" on it. It's important to know why plants should never be planted in the same space every year, and why gardeners must routinely fertilize their soil with nutritious options, like compost and manure.

What Happens to Soil After a Crop

Imagine if your fridge had three food items in it: spaghetti, pizza, and pot roast. Perhaps pot roast is your favorite, so every time you go to the fridge to find something to eat, you grab a plate of pot roast. What do you think would happen? *You would run out of pot roast.* Straightforward, right?

The same thing goes for soil. While you might not realize it, soil is filled with many different organic compounds that feed the plants. Each plant has different requirements and therefore absorbs different things out of the soil. If you grow plants in the same soil for too long, eventually it will deplete the soil of that particular nutrient, and they will stop growing.

Aside from depleting the soil of specific nutrients, plants can also expel specific nutrients into the soil. This happens as they break down and decompose back into the soil after the growing season. If you mix them into the soil, you can enrich the soil with plenty of great nutrients for other plants to grow. This way, your soil is never depleted, and you are able to continue growing in it for as long as you need.

The proper length to wait on putting a crop back into the same soil is four years. This means that you must rotate four different crops between one bed of soil to keep that soil as rich and nutritious as possible. As long as you maintain this cycle, your crops will always have everything they need to thrive.

How to Properly Rotate Crops in Raised Garden Beds

If you have a limited amount of space, the idea of crop rotation may seem challenging. That challenge may turn into frustration when you realize that your perfectly arranged garden will not be able to have that same arrangement for another four years. Rest assured, there are plenty of ways around this. The best way is by employing a garden layout where you break each garden bed into four sections. For example, if you have a garden bed that is 8ft long by 4ft wide, you would break it into sections of 2ft long by 4ft wide. You would then plant a different plant in each section for your growing season. The next year, you would move all of the plants one space to the right, then move the plant from the farthest right over to the farthest left side. This way, everything was fully moved over from where it was the previous year. This crop rotation system works excellent, whether you have one bed or multiple beds that you are growing in.

If you have enough raised garden beds that you are planting different plants in each container, you will simply switch which container each plant is in yearly. If, however, you cannot do this because it would result in tall plants being placed in the way of short plants, you should break your garden beds into four sections and follow the former advice to keep your crops properly rotated every year.

Chapter 5

Building Your Structures

After the planning comes the first step of action: building your raised garden bed structures. This part is easy, though it will require some planning and work of its own. You need to make sure that the structures you build are sturdy, that they will not rot from all of the water that will be contained inside of them, and that they are designed in a way that meets your needs. Several tools will be required for you to build these garden beds yourself. You could have someone else build your garden beds for you; however, that will be more expensive, so if you are looking to save costs, it is better to do it yourself. If you do not have all of the tools needed, look into renting them from local hardware stores to minimize your costs. This way, you can get your raised garden beds made, and you are not spending excessive amounts getting it done.

Preparing Your Garden Area

Before you begin any building efforts, start by preparing the area where your garden will be placed. Completely remove anything that is in the way of your garden beds, and if you can, remove anything that may be casting unnecessary shadows over your garden space.

Next, pinpoint where your raised beds will be placed and remove any grass from this part of your garden. Grass often has weeds mixed with it, and it can prevent proper drainage. Staking your garden beds into place is more challenging when grass is present. Use your shovel to remove the grass and toss it out of the way completely. Make sure you only remove the surface layer where the grass and its roots lie, and that you do not dig an actual hole into the ground. Once space is cleared of grass, mix the topsoil to loosen it up a bit. This will further improve the drainage for your garden.

Selecting Your Building Materials

Building your raised bed garden will require you to have the right tools to do the job. First, you are going to need materials that will be used to shape your garden beds. Concrete blocks are a great idea as they are permanent; however, they are more expensive. Galvanized tin is a popular option with raised bed gardens because it does not rot or rust. Still, it does have sharp edges that you will want to frame in to avoid getting cut by it when you are gardening. If you do use galvanized tin, be sure to use stainless steel fasteners so that your fasteners do not rust from exposure. Rot-resistant wood is a popular and inexpensive option, and it is relatively easy to work with. It will, however, break down over time, so you may have to do repairs on it every now and again.

If you are using rot-resistant wood, you will need two 2 x 6 x 8 pressure treated boards for the sides if you are making a 4ft x 4ft bed that is 6 inches high. You will also need one 2 x 4 x 8 pressure treated board for the stakes.

In addition to your building materials, you are going to need a selection of tools to help you get the job done right. These tools include: a tape measure, skill saw, drill and drill bits, screws or other fasteners, framing square, level, and a sledgehammer.

Building Your Garden Beds

If you are building a garden bed out of concrete blocks, all you will need to do is stack the blogs, so they are overlapping, reaching at least 6 inches high. You will need to create as many layers as are required based on the type of cement blocks you bought. If you are building your garden bed out of galvanized tin, you will need to have the tin precut and the proper tools to work with metal. If you are building your raised garden beds out of rot-resistant wood, you can follow the directions below.

The first step in building your raised garden beds is cutting your boards to length. You will do that by cutting one of your 2 x 6 boards in half so that you have two four foot lengths of board. You will cut the other one down to two 45 inch pieces. Lay your boards out on a flat surface as close to where the bed will be placed as possible. The two four foot pieces should cover the ends of the 45-inch pieces. Screw the longer boards into the ends of the shorter boards at each corner.

Place your built garden bed over the prepared plot you have made for it. Ensure that your garden is square and level, and adjust it if need be. Then, cut your 2 x 4 board into 4 x 18-inch stakes. Fix your stake on the inside corner of each of the four corners of your garden bed and use a sledgehammer to drive it into the ground until they are flush with the edges of your garden bed. This will keep your garden in place and prevent it from shifting. With raised beds made of wood boards, the weight of the soil and water can cause the beds to warp, and this prevents that.

Once your beds are prepared, you can fill them with soil and commence your garden plans per your growing chart.

Building Odd Shaped Structures

If your garden is going to be oddly structured, there is a different method that you will use to build your structure. For these beds, wood is the best option as it is the easiest to cut and shape to the odd shape of your garden space. To build your odd-shaped structures, you will use all of the same tools as above, except you will need to improvise on your boards and stakes to ensure that you get the right ones.

For odd-shaped gardens, round stakes are best because they are easiest to fasten boards from any angle. You will want as many stakes as you will have corners. If you are building a triangle, for example, you will need three stakes. If you have a particularly long edge, you may want an additional stake to reinforce that edge. Your stakes should always be long enough to reach from the top edge of your garden, down into the soil at 12 inches deep.

In addition to your stakes, you will need to measure out the edges of your oddly shaped garden and retrieve 2 x 6 boards in the required lengths. For example, if your triangle was 6ft on one side, 4ft on the other, and 4 ft on the last, you would need two 4ft boards and one 6ft board.

Once you have all of these tools, you will start by screwing the longest board to two stakes, with one stake on either end. Then, you will go to your prepared plot for the raised bed and place that board in its final space by driving the stakes into the ground. Next, you will drive your remaining stakes into the ground in their final locations. Ensure that they are placed in a position that will have them inside of your raised bed. The next board you fasten in place should share an edge with the original board that you have already placed. This way, you can screw it in without the first stake moving around. Once this board is screwed into the stake that is attached to the first board, you can screw it into the next stake on the other side. Continue doing this until all of your boards have been fixed to stakes.

After you have completed all of your boards, check your garden for levelness and use your sledgehammer to adjust it as needed, until you have a level garden. When you fill the oddly shaped garden with soil, be sure to spread soil into all of the corners to make the most out of your garden space. This will also prevent soil from falling apart and plants from falling over as the roots grow larger.

Vertical Gardening Structures

Vertical gardens are an excellent way to maximize your growing space and increase your yield. These gardens can be used in small backyards, side yards, and even on patios. Make sure that if you have multiple vertical gardens, they do not block each other's sunlight, as this would cause some of your plants to die.

When it comes to building vertical gardens, there are two excellent options you can choose from. Trellises and vertical cages are all great for growing vertical gardens. All three of these features can be built using simple materials at home, aside from trellises, which are easier to buy from the store. When you place trellises, you need to make sure that you fix them in place properly so that they are sturdy and able to bear the weight of the foods you grow on them. Since trellises do not carry soil, they can be used to grow crawling plants such as beans, grapes, and squash.

Vertical cage gardens can be built using 2 x 2s, a saw, a drill, and some 3-inch screws. To build a basic one, start by cutting twelve 20" lengths of 4 x 4, and four 48" lengths of 2 x 2. Lay four of your 20" lengths out flat in a square and screw them together. Make two more square with the other 2 x 2s. Next, you will lay one of your squares on a flat surface and hold one of the 48" lengths of 2 x 2 upright in the inside corner of the square. Screw it into place. Do the same in the remaining three corners of the square. Now, you are going to place another square over the four posts and slide it down until it is about halfway down the posts. Screw it into place. Finally, you want to fix the last square in place on the ends of the post and screw it in place so that the posts are flush with the edge of the square. This can then be placed over tomatoes or other large bushes that require added support for their heavy branches. The result is that the branches do not break off, and therefore they can bear fruit for longer. If you wanted to take this cage even further, you could place chain link fencing or chicken wire around it, fill it with soil, and plant herb seeds in the sides of it. Be sure to water it from the sides, and not the top, for the seeds.

While building your own cages is the best way to get the strongest vertical garden cages, there is a store-bought alternative you can consider. Wire vertical cages from garden stores can serve a similar job as these. However, they are not quite as strong or supportive, which means they may not give you the support you need in your plants. If they are all you can afford, though, they will offer an excellent alternative for your growing needs.

Creating Garden Covers

Garden covers are an important part of backyard gardening. Unlike in farmers' fields where large crops are grown, and so plants can protect each other, in small backyard gardens, the crop is usually quite a bit smaller. This means that things like wind, heavy rain, and other weather fronts can rapidly penetrate the plant and cause damage. Proper garden covers can protect your plants from damage by offering shelter if necessary. In some climates, garden covers may not be necessary. In others, however, they are essential to keeping your plants alive and thriving. Aside from helping protect your plants from exposure, garden covers can also allow you to plant crops earlier by creating a sort of temporary greenhouse around your plants.

The easiest way to create garden covers for your raised bed gardens is by using PVC piping. This will suffice for any square or rectangular garden. However, you can adapt this to support an irregular shaped garden, too. To start, you will want two 3-inch lengths of 1-inch solid PVC pipe per crossbeam you will be creating. Over a square garden, three crossbeams will be plenty so you will need six 1-inch solid PVC pipe pieces. For a rectangular garden, you will need one every two feet, so increase your crossbeams per two feet of garden space. This ensures that you will also have a crossbeam on both ends of the garden, too. This way, your cover goes side to side. You will also need tube straps that will hold your solid PVC in place. Next, you need to take lengths of ½-inch flexible PVC pipe. These will serve as your crossbeams. For a 4-foot-wide garden bed, you will need your PVC crossbeams to be about 6 feet long. Lastly, you will need a piece of material long and wide enough to cover your garden with. Heavy-duty polyethylene, polycarbonate, or polyvinyl chloride are perfect as they will all turn your raised bed gardens into mini-greenhouses. These greenhouses will be sufficient enough to protect your gardens from frosts occurring at temperatures as low as 22F.

To create your cover, you will start by using tube straps to fix your 1-inch solid pipes in place, where your crossbeams will lay. Start by fixing two in two corners on the same end of your raised bed. Then, every two feet after that, fix another one into place. You want to be crossing over the width of the bed, not the length of the bed. Be sure to use a square and a tape measure, so you are confident that the beams will be in the right place.

Once all of your solid PVC is in place, you can take the flexible PVC and fix it into place, too. To do this, simply insert one end into one solid PVC tube, then bend it over and insert the other end into the other side of the solid PVC tube, opposite the first one. Do this for all of the crossbeams.

The last thing to do is to fix your fabric in place. You can do this by throwing it over the length of your bed and pulling it across the width until it completely covers your garden bed. It can be fixed in place using heavy rocks, or other materials that will be heavy enough to hold it down. Avoid using screws or nails, as it will be challenging to open and close the fabric. Eventually, it will lead to tearing, which will lead to you needing to replace the fabric sooner.

Chapter 6

Managing Your Soil

Since your soil is responsible for providing adequate nutrients to your plants, it is important that you maintain your soil properly. Properly maintained soil will have plenty of nutrients available for your plants to consume, effectively creating the perfect growing conditions. Although soil generally comes packed with nutrients, there are many ways that you can improve the nutritional content of your soil so that there are more biological compounds available for your plants.

Where to Get Quality Soil

Filling your raised bed gardens is best done by purchasing additional soil to place in your raised beds, rather than using soil directly from your backyard. This freedom to purchase soil for your raised beds means that you can choose the best quality soil for your garden, effectively giving your plants the best chance at thriving.

Ensure that the topmost 2 to 8 inches of your soil have the highest concentration of organic matter and nutrients. As seedlings, this extra boost is necessary to help your plants grow strong and healthy. With larger plants, the main roots will remain in this space and will continue to absorb the nutrients out of this part of your soil.

The best place to look for high-quality soil includes garden centers and topsoil and mulch suppliers. Big box stores and resellers on public classified lists will have soil as well. However, the quality of their soil may be lower than dedicated garden centers and suppliers. Avoid getting your soil anywhere else because while it might look like typical soil to you, it will likely not have nearly as many nutrients in it as the soils from these centers will.

If you are on a tight budget, you can use the topsoil from your yard to help you reduce your costs. There are two ways that you can do this. One way is to use the topsoil from your yard in the bottom of your raised bed gardens and then purchase enough soil to fill only the top 2-8 inches. In this case, mix 75% compost and 25% topsoil from your yard for the base and then add the purchased soil on top. If you genuinely have no funds to devote to the soil, you can use the compost and topsoil mix for your entire garden. While this is not optimal, it will provide you with enough to get your plants going. When it comes to soil, though, you should always buy the best quality soil you can afford within your budget since it plays a significant role in the success of your crops.

Soil pH, Density, and Soil Mixes

Soil pH, density, and soil mixes all contribute to the quality of your soil, too. Soil pH refers to the acidic balance of your soil, density refers to how compacted your soil is, and mixes refer to what substances are mixed into your soil.

Soil pH matters because the acidity of your soil can impact the quality of your plants. In some cases, the wrong acidity can even kill your plants. Any pH below 7 is acidic, while any pH above 7 is alkaline. If your pH is 7 exactly, you have neutral acidity. Each plant will prefer a different pH range, though the majority of plants will thrive in a pH that is between 6.1 and 6.9.

Soil is more likely to become too acidic rather than too alkaline, so you will want to have correctors on hand that will increase the pH rating to decrease the acidity of your soil. The most popular corrector to use for your soil is limestone, which can be added to raise the pH to a more acceptable level of acidity. Wood ashes can also help correct soil by raising the pH.

Soil density refers to how compacted your soil is. Some types of soil, such as those with clay in them, are naturally denser. These types of soil are too dense for plants to reasonably grow in, as it is so compacted that their delicate roots cannot spread out and absorb nutrients. If you have soil that is too sandy, it will not hold water properly. Therefore your plants will not be able to absorb water quickly enough before it drains away. The best density of soil is a moist, fluffy soil that drains well so that it stays moist long enough for them to absorb nutrients, but not so moist that they never get a chance to dry out in between watering.

If you purchase your soil from a garden center or a topsoil or mulch supplier, chances are you will see options to purchase soil that is mixed with additives. These additives are designed to add certain benefits to the soil. Some additives support the balancing of pH, while others support the soil's ability to provide nutrients to your plants. Others still make the soil more fluffy so that it drains better. Purchasing pre-mixed soil is best because it breaks down to a lower price in the long run. As well, the soil has likely been sitting long enough that it has absorbed the additives, and they have done their job. Finally, purchasing pre-mixed soil ensures that you have the proper measurements of each of the additives that are placed in your soil.

Some of the additives you are likely to find in your soil include sphagnum peat moss, coir fiber, perlite, vermiculite, sand, limestone, fertilizers, composted wood chips, and compost. If organic gardening is important to you, ensure that you purchase organic soil. This ensures that there are no non-organic fertilizers or additives used that could leach into your plants and eliminate their organic factor. In organic soils, fertilizers are generally made out of composted materials and manure, rather than manufactured chemical fertilizers.

Maintaining Your Soil

Proper soil maintenance is important as it ensures that your high-quality soil stays high quality. Throughout the growing season, many things can occur that can damage the quality of your soil. Similar to the saying "you are what you eat," you can tell if a plant is not getting adequate nutrients for their needs because they begin to show signs of illness in their leaves. You can often witness poor quality soil in your plants through wilting, brown edges on the leaves, holes in the leaves, and yellowing of the overall plant. If this happens, you need to look into the quality of your soil, as well as at the possibility of there being pests that have infected your plants.

If pests have infected your plants, you will find them by turning the leaves over and inspecting the bottoms. Inspect several leaves on each affected plant so that you can locate all of the bugs that may be causing issues. If you spot them, you can remove them from the plants and move them far away from your garden. If they are too small to remove, you will need to identify what types of bugs they are and then use insecticidal soap to remove them from your plants. These soaps are designed to remove pests and insects without harming your plant, and they often work best when you spot the problem sooner rather than later. Check your plants regularly for signs of infection and treat your plants as soon as you spot anything that could be harming them.

If soil is your issue, you can fix that issue by properly maintaining your soil. Checking your soil for pH, as well as for nitrogen, potassium, and phosphorus is important, as these are all essential to a plant's well-being. These tests are often called pH tests and NPK tests and can be found at most garden centers. Removing debris and weeds from your soil and adding mulch to the top of it can help, too. As well, keep an eye on your plants and do anything you can to help them thrive. If you see a plant leaning over because it is too heavy from the fruits or vegetables, it is bearing, put a cage around it or attach it to a trellis to take some of the pressure off of its branches. During heavy rain, cover your garden if your plants have not outgrown your covers.

If they have, keep an eye on your soil and do not water it again until it has dried out from the heavy rains.

Utilizing the Power of Composting

Compost is essential to gardening because it provides the garden with necessary organic matter and nutrients for fruits and vegetables to grow. Compost is added as a mix to your soil to improve your soil structure, effectively giving your plants a better base to grow out of. Fortunately, you do not have to rely on buying expensive compost mixes, either. You can make it in your own backyard out of things you have in your household!

To start your compost, you want to place it directly on bare earth. This way, worms and beneficial organisms from the ground can get into your compost to aerate it. Keeping your compost aerated is important as it prevents it from becoming too wet and soggy. The rest of composting has to do with layering proper materials into your compost and regularly turning the compost so that it dries out, and so that all of the layers crumble together.

On the bottom of your compost, you want to lay a few inches worth of twigs or straw. This layer also provides aeration by giving your compost adequate drainage so that the soil does not stay moist for too long.

For the remaining layers of your compost, you want to alternate between dry and moist layers. In dry layers, add straw, leaves, sawdust pellets, wood ashes, and even dry grass clippings from your lawn. In moist layers, add tea bags, food scraps, seaweed, and any other moist additives you might have for your compost. Once you have created a few dry and moist layers, you can layer some manure into your compost pile. Manure offers a rich nitrogen source that activates your compost pile and helps accelerate the process for turning it into a rich soil additive.

The compost should be kept moist, but not wet. Every so often, when the compost is looking dry, add some water to it or uncover it so that some rainwater can naturally irrigate your compost. The water will absorb the nutrients in the soil and move it around, while also allowing it to grow even richer.

Covering your compost is important as it keeps it from drying out or getting too wet if you are in a climate that gets a lot of rain. The covers will also help your compost stay warm. Warmth will increase the humidity of your compost, which further promotes the development of various nutrients in the soil. Wood and plastic sheeting are both great materials to cover your compost.

Every few weeks, use a pitchfork or a shovel to turn the compost. Turning your compost adds oxygen, while also helping combine everything in the soil. After you have turned your compost, resume adding in your layers and following all of the same steps to develop and maintain your compost.

You will know your compost is "done" when it has a crumbly, smooth texture, and you do not see any remnants of the food items or other additives you have placed in your compost. You might see some woody or fibrous pieces, or some avocado pits or corncobs, however, nothing else should be recognizable. Peelings, leaves, tea bags, and other such items should all be completely broken down and incorporated into the compost. Two other things you will notice when your compost is done include its scent and its color. The scent of compost should be sweetly fragrant and loamy. If you notice any sour odors or ammonia-like odors, your compost will need longer to mature. The color of your compost should be a nice, dark color. You will likely notice that it almost looks black because of how much has been added to it. This is how compost has earned the title "black gold."

It will take your compost anywhere from four weeks to twelve months to be finished. The factors that affect the rate at which your compost is complete include things like: the size and type of organic matter added to your compost, how often you turn the pile, and whether you are using a hot or cold method. Chopped and shredded organic material will compost faster, and turning it frequently ensures that you are allowing it to decompose faster with the support of aeration. Hot composting methods will break foods down faster, too, while cold composting will take longer.

Using a compost tumbler is a great way to create compost rather quickly. Compost tumblers allow you to pour everything in, keep the compost environment humid and warm, and simply turn the bucket to aerate it properly. With a compost tumbler, you can get completed compost in as little as three weeks.

Once your compost is mature, you can mix it with the topsoil in your garden to provide it with better nutrients for the plants that you will be growing there. If you are already into the growing season and your plants are starting to mature, or have completely matured, you can use compost around the surface of your garden as a sort of mulch. This will allow the soil to absorb more nutrients and may even extend your gardening season to provide for a longer harvest.

Chapter 7

Planting Your Garden

When all of your conditions are just right, it is time to start planting your garden. Planting your garden requires a certain skill, as you need to know that your plants are properly spaced, in the right location, and that they have access to what they will need to thrive. Some of your plants may not be strong enough to be planted directly into the gardens, so you might have to start some indoors with small greenhouses so that you can give those plants a better start at life. Once you have turned them into thriving seedlings, you can transplant them into your garden beds so that they can live there for the rest of the season.

Properly planting your seeds and seedlings will require you to get the timing and technique right. It is important that you follow your growing chart as closely as possible to ensure that you can get as many plants going as you can.

Understand that just because you plant a specific number of seeds does not mean that you are going to end up with that many plants. Some of your seeds will fail to grow, others will die as seedlings, and others still may fail to thrive after reaching a certain growing point. For that reason, you need to be prepared to take quick action if some of your seeds do not work out so that you can still obtain the maximum yield from your crops.

Buying the Best Quality Seeds

Seeds can be found in every garden center and most big box stores during the gardening season. In big box stores, they come into stock in the late winter and early fall and typically stay until they are sold out. With all of the different places to buy seeds and all of the different types of seeds, it can be easy to get overwhelmed. Fortunately, there are a few things you can do to ensure that you are getting the best quality that will also provide you with the best quality plants.

To purchase seeds, you can go directly to a store and look at their supply, or you can use a seed catalog to help you find what you are looking for. Seed catalogs can be found both in store and online, providing you with many options to acquire the seeds you need for your garden. Keep in mind that if you order your seeds online, you may end up paying extra for shipping expenses, so factor that into your budget.

By this point, you likely already know which types of plants you are going to be planting in your garden. However, you may not know which subspecies of that plant you will be planting, yet. The best way to determine subspecies is to ask questions and to consider your preferences and needs. Start by asking local gardeners and farmers which subspecies they have used, and which company they bought their seeds from. This will give you an idea of what types of subspecies thrive in your local area and where you can buy those seeds. Once you have a good idea of which subspecies will be best for your area, you need to decide which will be best for your garden and family. Choose a subspecies that will thrive in your garden's conditions, and that will give you what you are looking for. For example, if you prefer long, thick carrots over thing finger-thin carrots, choose a species that is known for growing longer and thicker. This way, you are growing exactly what you want.

On seeds, you will find labels as to what the seeds are, or where they come from. Four of these labels are good and are ones you should look for, while three are ones that you should avoid in your garden. Heirloom, organic, open-pollinated, and conventional and naturally grown seeds are all great options to choose from. Hybrids, genetically modified seeds, and treated seeds are all ones you should avoid buying.

Heirloom Seeds

When an open-pollinated plant goes to seed, those seeds are labeled "heirloom seeds." Heirloom seeds have been passed down through generations. They are saved because they have a specific flavor that is preferred, they are productive, they are often drought-tolerant or hardy, and they have great adaptability. Heirloom plants date back at least 50 years, though some date back as long as 300 years. If you would like to preserve cultural heritage through your garden, heirloom seeds are a great way to go about it. There are a wide variety of heirloom seeds for each type of plant, so you are still going to need to decide on which subspecies you want to grow so that you get exactly what you are looking for.

Organic Seeds

Organic seeds must be certified organic. These seeds are cultivated using only organic gardening methods. In the United States, organic seeds have a USDA organic label which guarantees their organic quality. If you grow organic seeds, you know that right from the seed the plant you are growing is truly organic. Aside from purchasing organic seeds, you must keep up with organic gardening methods and eliminate the concerns of using harsh chemicals anywhere around your plants.

Open-Pollinated Seeds

Heirloom seeds are always open-pollinated, but open-pollinated seeds are not always heirloom, nor are they always organic. The benefit of open-pollinated seeds is that you are going to get a seed that will grow to be exactly like the parent plant it came from, assuming that no cross-pollination happened during the process. The benefit here is that you end up with plants that have a familiar great taste and that they will last for seasons to come.

Conventional and Naturally Grown Seeds

Conventional and naturally grown seeds are seeds that are not certified organic. They are often the cheapest seeds you can buy, and they are often found in box stores or hardware stores that have a garden department where they are selling seeds from. While these seeds will give you great plants, it is important to note that they may not give you the best quality of plants because of how they are grown and harvested. Note that any plant you grow from these seeds will not be truly organic, even if you use organic growing methods, because they were harvested using conventional methods. These seeds should be a last resort option if they are all you can access or all you can afford. Otherwise, opt for heirloom, organic, or open-pollinated seeds.

Hybrid Seeds

Hybrid seeds are created by crossing two varieties of plants together based on the desirable traits either possesses. The idea is that you will receive a plant that provides you with the benefits of the two plants that were combined to create the new plant. For example, maybe one plant bore larger fruit while the other had a longer shelf life. By combining the two, you get larger fruit and longer shelf life from your seeds. Hybrid seeds are labeled "F1," and they are not the same as open-pollinated seeds that are labeled "OP." Open-pollinated seeds experience cross-pollination that occurs naturally, while hybrid seeds are man-made.

While hybrids offer benefits such as disease resistance, improved flavor, faster maturity, increased productivity and higher yield, easier storage, and uniform growth, they are not necessarily the best for you to purchase. Hybrids are not "true to type" like open-pollinated seeds are. Sometimes, they are not viable. Other times, the plant may revert to being similar to just one of the parent plants instead of retaining both of their characteristics. Further, the seeds you harvest from this plant are unlikely to give you the benefits of the hybrid seeds. The lack of viability mixed with the unreliability of these seeds makes them unsustainable for most gardeners.

Genetically Modified Seeds

Genetically modified seeds are bio-engineered seeds, and they should be entirely avoided. These seeds take the hybrid process a step further by splicing together two seeds from two entirely different "kingdoms" to create new flavors, shapes, textures, and other things in plants. This means that they are created through man-made processes just like hybrids are, only they are done so in a more intense way. While hybrids are cross-pollinated on purpose, genetically modified seeds are actually created in a lab.

In recent years, it has been proven that genetically modified seeds and genetically modified plants are capable of creating significant health issues in humans. In fact, these health issues are not only in humans. Even the insects who tend to eat these crops have been dying off as a result of consuming them, proving how dangerous they can be.

Most plants do not have a genetically modified option, especially not those that you will be growing in your garden. There are, however, a few varieties that you are likely to come across that will have genetically modified seed options. These include corn, sugar beets, potatoes, soybeans, and alfalfa. If you were planning on growing any of these plants, be sure to select a variety that is not genetically modified.

Treated Seeds

Treated seeds are seeds that come in unusual colors. Pink pea seeds, green cucumber seeds, and blue sunflower seeds are a good example of treated seeds. Seeds are treated with a fungicide, which is said to protect them from seed-borne and soil-borne pathogens. The problem is, these can deteriorate the quality of your soil and all of the beneficial living organisms inside of your soil.

Toxic chemicals are not required to grow your food. You can opt, instead, for natural methods for treating issues in your crops so that you are not poisoning your soil and creating problems for yourself down the line.

Spacing Requirements for Plants

Each plant you decide to place in your garden is going to have different spacing requirements. Adequate spacing ensures that your seeds have room to grow. This room provides them with an abundance of space to set their roots, as well as for their branches and leaves to grow into. It is important that your plants have some space between them so that when they are mature, they are not touching each other too much. This way, air, and light can penetrate through your plants and support them in completing their digestive process.

In raised bed gardens, spacing requirements are different because you are not using conventional gardening practices. The optimal soil conditions mean that plants can grow down rather than out, which results in their roots having more space without you having to put so much physical space between the seeds. With that being said, there are still requirements you need to consider.

To find information regarding proper spacing for the plants you are planting, look at the seed packet or seed catalog that you have for the seeds you are using. These places will contain plenty of great information on how to plant your seeds properly. Plants that are going to grow above soil should have more space than plants that are known for growing below the soil, as they are the ones that will continue to have larger space requirements. With that being said, be mindful of plants that are known for growing down into the soil and growing outward, as they will require more space, too. For example, potatoes tend to grow out quite a ways so they will need more space than, say, a carrot which only grows down.

For your plants that grow in the ground, consider how much room they are likely to take and plant them accordingly. Garlic, carrots, and onions, for example, can all be planted within' 3- to 5-inches of each other because this will give them plenty of room to flourish. Potatoes, on the other hand, should have about 12 inches of space between them so that they have plenty of room to flourish.

Plants that grow above the ground should always have at least 12-inches of space between them unless they are known for taking up more space. In this case, you will want to give them closer to 15- to 18-inches of space. Pepper plants and larger tomato plants are a great example of plants that will require more space for you to allow them to thrive. If you are growing plants such as squash, which will need to crawl in order to thrive, you need to be sure that you have adequate room for them to crawl. If you have planted them in a bed with carrots and beans, they can crawl all between the stalks. If not, you will need to find somewhere else for them to grow into.

Planting Them in the Proper Location

As you get into the planting process, you need to make sure that you are planting your seeds in the proper location. By now, you have already planned out where each crop is going to go in your garden. However, you still need to be sure that you are planting those seeds exactly where they need to be so that they can thrive in that location for the rest of the season. Some seeds will need to be planted in rows, while other seeds will need to be planted in blocks. These planting methods will determine exactly which location your seeds should be planted in so that you can get them exactly where they need to be. Once a plant has been placed into your raised bed garden, you do not want to have to move them. Moving plants around is not only a significant hassle, but it can also lead to you damaging the roots or disrupting the plant to the point where it dies. As soon as your plant is placed in the soil of your raised bed garden, you need to leave it there and tend to it in that location. If you cannot plant something directly into your garden, you will want to start your seeds indoors or in a proper greenhouse, first. We will talk more about starting seeds later.

How to Plant Seeds in a Row

Sowing your seeds in a row makes it easy for you to access all of the seeds you have planted. It is also a great way to plant companion crops as it allows you to plant two or three rows side by side, one for each plant in the companion crop. Farmers grow their crops in rows because machines can move down rows more easily, which means caring for and harvesting the seeds becomes easier, too.

When you grow your seeds in rows, it is important to ensure that the growing conditions are uniformed across the row. Ensure that there is uniformed light, shade, moisture, and access to nutrients in the soil across the length of the row so that your plants all have the best opportunity to thrive. If you do not have the ability to offer that uniformed growth opportunity, you are going to want to plant in blocks instead.

To plant your seeds in rows, you will start by planting two stakes, one at either end of the row you want to plant. Then, you will tie a piece of twine between the two stakes so that it clearly marks the row for you. Now, you will use a trowel, a furrow, or your finger to carve out a line along your row. The seed packet of the seeds you are planting will tell you how deep that line should be. Once you have carved out the line, place a tape measure alongside the row so that you can measure the distance between your seeds. Place your seeds along the row, then remove the tape measure and the row markers. Lightly cover your seeds in soil by pushing the dirt back over the seeds. Do not pack the soil down, though, as this will make it too dense for the seeds to grow through.

How to Plant Seeds in a Block

If you have decided to split your garden beds up into sections, you are going to need to plant your seeds in blocks. Sections, or blocks, are great for improving your ability to complete proper crop rotation, to ensure that each plant has access to the proper conditions it needs, and for organizing your plants.

You can plant your seeds in blocks by first creating a row marker along the sides of your section so that you can clearly see where that section starts and ends. With the required measurements in mind, use a trowel, a furrow, or your hand to carve out a square shape inside of the block. So, if you need 4-inches of seed spacing for the crop you are planting, you will make that square 4-inches away from all edges of the section you are planting. Once you have, you will use a tape measure along the trench you have dug out to place your seeds properly. Place them as deep as is recommended on the package or catalog for the seeds you are planting.

After you have finished the first block of plants, you will use your tape measure to measure in another four inches from the block you have just planted. Then, you will dig out another square and repeat the process. Continue doing this until you have completely planted the soil with as many plants as you reasonably can. Then, move on to the next block of your soil to get started with a different crop.

Giving Plants What They Need to Thrive

Planting your seeds in the garden is only the first step of starting your seeds this way. Once they are planted, you need to continue caring for them properly so that they can grow properly and thrive in your garden. At this point, you should have already filled your garden with high-quality soil and compost, so the only thing left to do is keep your seeds moist and warm.

Seeds have different watering requirements than mature plants because they do not yet have a root network to absorb water through. Instead, they absorb water directly through their shell and grow from there. Too much water can drown your seeds and prevent them from properly growing. Ideally, you should sprinkle your seeds with water each night so that the soil never dries out. If you find that your environment is exposing your seeds to harsh growing conditions, cover your garden beds with the garden covers you have created for them. This will ensure that they remain humid, moist, and protected from the elements until they are growing steadily.

Always be patient with your seeds, as it can take anywhere from a couple of days to a week or two to get them really going. As well, it can take a while for them to mature from seedlings into proper little plants. Continue to focus on watering them and keeping them moist and protected from excessive temperatures, excessive moisture, or excessive dryness so that they have everything they need to thrive. Then, wait for them to grow!

Avoid fertilizing your seeds or adding anything to the soil once you have planted them. Doing so can disrupt the seeds and introduce harsh chemicals to the soil, which will then be absorbed by the seeds, too.

Starting Seeds Indoors

Starting your seeds indoors is a great way to be able to start your seeds as early as possible, effectively extending your growing season so that you can yield a higher harvest. Every single seed you would plant in your garden can be started indoors. However, many people will start vegetables that grow underground directly in their garden to avoid disrupting the roots of that plant.

Although starting seeds indoors may take up a lot of space, it does not take a lot of effort to get started. Growing trays can be purchased, which allows you the opportunity to add one seed per section, making it easy for you to reach the necessary spacing requirements for each plant. You can purchase growing trays that have pods of soil already inside of them, or you can plant growing trays with your own pods of soil. Generally, it is cheaper to plant them with your own soil so that you do not have to repurchase the pods every year.

Once you have arranged your trays with soil, all you need to do is add one seed per section. As soon as everything is planted, mist them with some water so that they have everything they need to start growing right away. For indoor seeds, a misting bottle works best for watering as it prevents you from penetrating the soil with heavy droplets of water or drenching the soil with streams of water. Some growing trays will have plastic covers you can use to cover the tray so that they stay moist and warm. If you do not have plastic covers for your trays, use plastic material to cover them so that you can create that warmth inside of your trays. Stop covering your trays with plastic when the seedlings start to touch the plastic; otherwise, it will prevent them from growing any further.

When it comes to placing your seeds in the house, you want to ensure that they have ample access to sunlight. However, you do not want them to bake in the heat. Placing seedlings directly in windows can result in them becoming too hot and dying. Instead, place them about 3-4 feet away from the window, but in direct light from the window itself. This way, they get ample light, but they are not getting baked by the sun.

Planting Seedlings in Your Garden

There are a few processes that need to be completed before you can transplant your seedlings outdoors. The two biggest practices you need to be aware of are planning and hardening off your seeds. Proper planning, and time spent preparing your seeds for outdoor conditions, ensures that your plants will be ready to take to the garden effectively. This way, they continue to thrive even.

Knowing When to Plant Seedlings

Planning ahead is imperative when it comes to transplanting your seeds outdoors. Improper timing can result in you losing your entire crop, so you must ensure that you are following proper practices to succeed. If you transplant your seedlings too early, frost can destroy them. If you plant them too late, they may get baked in the sun because they did not have adequate time to get used to being out in the heat before the weather got significantly hotter. Paying attention to your local weather conditions and planning ahead ensures that you can begin the hardening off process on time and that you transplant your seeds at exactly the right time.

The first step in planning your transplant is checking your local planting calendar to see when the last spring frost should be in your area. This date is generally a guideline, not a hard and fast rule, so you will want to continue to pay attention to live weather updates, too. However, this will give you a period to aim toward which will help you decide when to start hardening off your seedlings, and when to transplant them officially.

Specific plants will be more likely to thrive in cooler conditions, while others will not be. You should identify which your crops prefer and adjust your plan accordingly. Cooler crops such as spinach or peas should be transferred out earlier before temperatures rise too high. Other plants that prefer warmth, such as tomatoes and peppers, should be transplanted later on as they are not likely to survive a frost. For warm-weather crops, keep a close eye on your local weather forecast and avoid planting them until you are certain that there are no more frosts or cold weather snaps in the forecast. You can identify which crops are cool weather and which are warm weather by simply looking up the crops you are growing. Once you have that information, keep track of it so that you can harden off your cool weather crops earlier, and you can harden off your warm-weather crops later.

As you plan all of this out, be sure to keep track of everything in your plant journal. Write down when you started your seeds, and when you transplanted them, as well as how it went for you. This way, in future years, you have a general idea as to when you can start hardening off your plants and when transplanting should occur. Over time, you will discover that this helps you create even better success with your indoor-started seedlings.

Preparing Your Garden and Plants

Preparing your garden for your seedlings requires you to intentionally warm up the soil, as well as loosen it up in case it has hardened through the winter. If your garden beds are brand new this year, there will not be much for you to do, as you have already largely prepared them. The best way to ensure they are ready for your seedlings is to apply your garden covers so that the temperature of the soil remains warm, as this will make it easier for the plants to transition from the warm indoors to the cooler outdoors.

If your raised beds are older, there are a few things you will need to do to prepare them for the incoming seedlings. Turning the soil and adding some finished compost is a great way to ensure that your seedlings have plenty of soft, loose soil and nutrients to help them make the transition. Covering the garden in your garden covers will help, too.

Preparing your seedlings for outdoors requires you to engage in a process known as hardening off your seedlings. Hardening off your seedlings starts by withholding fertilizer and reducing the amount of water you give your seedlings, as this prepares them for what life will be like outdoors. Since you will only be watering them once, maybe twice per day, reducing the frequency of watering helps them get used to this. Avoid cutting back too quickly, and instead reduce the watering day by day to prepare them for this change.

The other part of hardening off your seedlings requires you to expose them to the outdoor environment in a slow, sheltered manner. You can do this by placing them outdoors in a sheltered area 7-10 days before you will be transplanting them. This area should have dappled shade, such as under a leafy tree, and should be protected from the wind and other elements. Set them out for a few hours each day, gradually increasing their exposure to the elements each time. While they are outdoors, ensure their soil remains moist at all times, as dry air and spring breezes can lead to rapid water loss, which can shock and kill your seedlings.

Since you will be transplanting your seedlings into raised, covered garden beds, you can use those to support you in preparing your seedlings as well. After about five days of hardening off, start placing your seedling trays directly into your raised bed gardens on top of the soil. By day 7-10, you should be able to transplant your seedlings directly into the soil of your raised garden beds.

Keep your garden covers on for the majority of the time, removing them only during the more mild hours of the day at first. This typically falls in the late morning, before noon. Be sure to cover them up overnight to help them remain warm and moist, and to protect them during the night hours. Within a week or two, your plants should be hardened off enough and strong enough to endure life without covers.

Once you remove covers full-time, you will still want to keep them handy on certain occasions. So long as your plants remain shorter than the covers, covers can be used to minimize exposure to heavy rainfall, wind, and other elements that could damage your young plants. By the time they are large enough that they no longer fit into the covers, you can stop using the covers altogether as they should be strong enough to withstand any weather that comes their way.

Chapter 8

Growing Your Garden

You might think that after you have planted your seedlings into the garden and started to see some growth out of them, the hard part is out of the way. Gardens continue to require diligent maintenance throughout their entire growth cycle to ensure that they are growing properly. Knowing how to maintain your garden ensures that you keep the growing conditions exact, while also preventing unwanted pests or illnesses from taking over your plants. It takes some time to get used to maintaining your garden, but once you do, it feels incredibly simple. These tasks are not challenging or labor-intensive, depending on the size of your garden. Still, they do need to be completed regularly to preserve the quality of your garden.

To give you an idea of what your time and energy investment will look like, you need to be out in your garden performing maintenance tasks every single day. Daily, you need to water your garden, pull weeds, inspect your plants for bugs, and perform general maintenance on your plants. Occasionally, you will also have to thin out seedlings, mulch your garden beds, fertilize your growing plants, and possibly perform more substantial weeding sessions if your daily ones are not keeping up with the weeds.

Watering Your Garden

Because water is an essential element to plant health, you might be surprised to learn that less is actually more. Overwatering your plants can lead to water-logged roots, causing a variety of diseases in the roots of your plants. Rather than overwatering your plants, you should aim to slightly underwater them, as this prevents things like root rot or the development of mold in your soil. The only time overwatering is preferred is if you live in extreme drought conditions, in which case your plants will have a tendency to dry out too quickly, which prevents them from absorbing nutrients fast enough. If you are in drought conditions, watering your plants twice per day and overwatering them is ideal.

Watering should be done in the early morning, though you can do it later in the evening around late sunset if you prefer. These two hours are best because the ground is not too hot, and the sun is not too hot, either. This means the water will not warm up and bake the roots of your plants or cause sunburn on the leaves of your plants. Both of these conditions can severely damage your plants, and even kill them in some circumstances.

Some plants will go limp during the middle of the day. You might think this is because they are dehydrated and believe that it is a good idea to water them right away. While they may be dehydrated, this is not a good idea. Even a limp plant can experience sunburn and burnt roots from watering during the hottest parts of the day. Instead, wait until the evening when the sun cools off and then water your plant. This way, it receives hydration, but not at the risk of developing sunburn or burnt roots.

If you come from a particularly damp environment, or if it recently rained, you will need to check to see if your plants require water before giving them any. This way, you refrain from overwatering them. To check your plants, place your finger into the soil in several different areas of your raised beds. If the soil is dry at two inches deep or less, you know you need to water your plants. If it is still wet, you do not need to water them yet.

When it comes to vegetable gardens, you should never rely on sprinklers to maintain your garden for you. While sprinklers are great for lawns or massive farmer's fields where hand-watering is not possible, they are not ideal for backyard gardens. Sprinklers deliver uneven water to the soil and can lead to certain plants being overwatered, while others are missed completely. Watering your plants with a hose using a proper garden attachment, or using a watering bucket are the best ways to ensure that water is delivered evenly across your soil and to each of your plants.

Thinning Out Seedlings

Thinning out seedlings is a method used to ensure that your plants have enough room to grow as they mature. How, and when, you thin out your seedlings will depend on where you started your seedlings, and how many are thriving.

If you started your seedlings in the garden and they are all thriving, you will cull any that appear as if they are thriving less than others. Only leave one thriving plant in each planting position to ensure that seedlings do not begin to choke each other out as they compete for water and nutrients in the soil. It may seem wrong to pull out and toss plants that are otherwise thriving, but if you don't, both seedlings will die. If only some of your seedlings are thriving, you can carefully pull them and re-plant those which are thriving in areas where more plants can be placed. Ensure you continue to follow proper distancing measurements for that particular plant type so that, as they mature, they have room to grow without competing with each other.

If you started your plants indoors, you can thin your seedlings out while you transplant them. To do this, take one potted seedling and carefully use your hands to pull the two seedlings apart, taking care not to rip their roots or damage the plants in the process. Plant each seedling in its long-term position in the garden. If you have too many, you can plant some in their own small planters, or you can toss them in your compost pile.

Mulching Your Garden Beds

Mulching garden beds are a great way to retain moisture and regulate temperature. Mulch also prevents weeds from going by creating a thick, impenetrable layer that is challenging for weeds to grow through. You can purchase mulch products from any garden store or garden center in a big box store, or you can use things from your yard like shredded leaves or straw as a natural mulch.

You should not mulch a garden that has seedlings in it, as the mulch will provide a heavy layer for the seedlings to grow around. Instead, wait until the seedlings have developed into hardier, more mature plants. This way, they are more durable and stand tall enough above the soil that they won't accidentally become buried by the mulch. At that point, you can place a layer of mulch 4" deep around your plants, coming up close to the base of each plant but not touching it. This will give you all of the benefits of mulch, without risking the lives of your plants.

Fertilizing Your Growing Plants

Some gardeners love using fertilizers, while others prefer not to. When considering whether to fertilize or not, realize that there are many options you can go with if you do choose to fertilize. Many believe that using fertilizer automatically means that you are covering your plants in harsh, unwanted chemicals, but that is not the case. These days, many natural, organic fertilizers exist that feature no harsh chemicals. Instead, they simply add to the nutritional profile of the soil to improve the growing conditions for your plants.

I strongly advise you to use fertilizer, though you can choose which fertilize you think will be best for your garden. This fertilizer will ensure that your plants have everything they need to thrive, improving your likelihood of having strong plants that provide an abundance of vegetables for you throughout your growing season.

If you choose to fertilize, find a good all-purpose or general-purpose plant food that can be used on virtually all of your plants. This way, you are not left trying to mix, manage, and distribute several different types of fertilizer. Fertilizers should be applied once every two weeks, as this is about how long it will take for the plant to consume those additives from the soil.

Weeding Your Garden Beds

Weeds that grow in your garden will compete with your plants, stealing nutrients and space away from them. Too many weeds can essentially drown out your vegetable plants, resulting in them dying off, or struggling to produce a sufficient harvest. Further, weeds that are left too long will go to seed, causing even more weeds to grow around them. You must remove weeds promptly to avoid having them become a bigger nuisance to your garden.

While mulch will go a long way in helping minimize the growth of weeds, it will not eliminate the growth completely. Weeds will continue to grow through your soil, and they will need to be removed. The best way to keep up with weeds is to look for them before watering your garden beds, and removing any you see. This way, you are keeping up with them before they become a major issue.

Resist the urge to leave weeding for the weekend or to put it off until a later day, as they can multiply rapidly, and then they become a large chore to manage. Not to mention, the more weeds there are, the more seeds can be distributed throughout your garden, which can lead to even more weed issues going forward. Anytime you spot something in your garden that should not be there, remove it.

Weeds should always be tossed directly into your compost bin after being removed. This way, they are no longer present around your garden and cannot pose a threat to your garden's health anymore. Never leave a small pile of weeds in the corner of your garden, or too close to your garden in general, to "deal with later," as this can create a larger problem within a matter of days. Many new gardeners become overwhelmed by the number of weeds they come across, which makes gardening seem more challenging than it needs to be. Regular maintenance is the best way to avoid this.

Inspecting Plants for Bugs

Bugs, like weeds, can become a serious threat to your plants. Bugs can eat your plants, some of them to the point that your plant dies. Others will introduce illnesses to your plants, or render your plants inedible because of how much they have damaged them. Dealing with bugs can be a nuisance, as they can deplete your harvest completely, and many can do it relatively quickly.

There are three types of bugs you need to look out for in your garden. Individual bugs, bug communes, and bug eggs should all be removed from your garden as soon as possible. Individual bugs like snails, caterpillars, moths, and certain worms will consume the leaves of your plants, and some will even burrow into the fruits and vegetables so that they are no longer edible for you. Removing them prevents them from being able to do any damage in your garden. Bug communes are generally bugs like aphids, maggots, and potato beetles, and they will group together on your plants and deplete an entire plant of its nutrients before moving on to the next one. Bug eggs, such as those belonging to the larva of the aforementioned bugs, can also cause damage as they use the nutrients of the plant to grow. As well, once all those bugs hatch, they will contaminate your plants and cause damage.

Like with weeds, you need to be inspecting your garden for bugs every single day. This should also be done before watering, as it makes inspecting your plants much easier. To inspect your plants, you want to look all over them and around them visually. You also want to turn their leaves over to inspect the bottoms of their leaves, as eggs and bug communes will generally live underneath leaves where there is shade from the hot sun, and from the other elements.

If you spot an individual bug on one of your plants, you can simply pull that bug off and toss it away from your garden. If you spot a commune of bugs, or a nest of bug eggs for bugs that are known to be damaging to your plants, such as potato beetles or maggots, you should clip that leaf off and boil it before tossing it in the garbage. This removes the entire nest of eggs from your plant and kills the eggs so that they cannot infest your garden any further. If you have an infestation across your plants, you can get a high-quality insecticide and "wash" your plants in them, as this should kill off any bugs and their eggs and larva so that you no longer have to worry about them destroying your plants.

Not all bugs are bad for your garden, so be sure to properly ID bugs before you decide to do anything about them. Good bugs should be left alone, as they will actually improve the living conditions for your plants. Many will even eat the bad bugs from your garden, providing you with natural pest protection for your plants. Some good bugs you want to keep around include ladybugs, praying mantis, spiders, ground beetles, aphid midges, braconid wasps, damsel bugs, and green lacewings. Bumblebees and earthworms are good for your garden, too, though grub worms should be removed as they will eat the roots of your vegetables, causing them to die.

Maintaining Your Plants

Each day, inspecting your plants for general health is an important step in maintaining a high-quality garden. Inspecting your plants allows you to remove dead leaves or branches, discover possible illnesses before they get out of control, support your plants in growing stronger harvests, and to identify the exact moment a plant is ready to be harvested. As you weed your garden and inspect your plants for bugs, you can also be inspecting them for general health.

Removing dead leaves and any other decaying debris from your plants should be done as soon as any is witnessed, as this prevents the development of diseases in your garden. Once diseases develop, it can be challenging to maintain them or prevent their spread, so this is important.

Each plant you grow will have certain things you can do to support it with growing stronger and healthier. Research your chosen crops, so you know what tasks to perform to improve harvest. Certain plants, like basil, will require you to pinch off flowers, or cut stems down shorter to encourage two stems to grow where one previously existed. Others, like tomatoes, may require you to prop them up with support so they can grow without breaking or experiencing other forms of damage. Providing this support as it is needed encourages your plants to have a strong, healthy harvest every single year.

Anytime you notice a plant seems to be preparing for harvest, you should immediately plan for and begin the harvesting process. Harvesting is something that should never be put off, as vegetables will rapidly accelerate past their prime and either be consumed by pests or will rot before you get a chance to pick them. Picking them sooner, rather than later, ensures the best quality harvest. We will talk more about harvesting your plants in the next chapter.

Chapter 9

Harvesting Your Crops

Harvesting your crops must be done at exactly the right time to ensure that the foods you harvest are ready and that they are not taken over by pests or rotten before you get to them. A fruit or vegetable that is harvested at exactly the right time will contain peak levels of beneficial nutrients and vitamins, while also being at peak freshness, making it best for eating or preserving. It can take some practice to identify exactly when your plants are ready for harvest, but there are many things you can do to prepare yourself for what to look for and what to expect.

Checking for Plants That Are Ready

Every day, as you work in your garden, you should be checking to see if any of your plants are ready for harvesting. Keeping an eye on your plants each day ensures that you are ready to harvest produce at exactly the right moment, so it is best for consumption. Be sure to look in unexpected areas, too, as some fruits and vegetables will be hidden under leaves or branches and may be hard to spot. Squash, for example, has a tendency to remain hidden under the big round leaves and may go unnoticed until it is too late. Ideally, you want to harvest everything you possibly can off your plants, as this maximizes your yield and encourages your plants to produce more.

Once you begin to see signs of produce growing on your plants, you know it is time for you to pay close attention to harvest time. You will want to look up signs of readiness for each unique plant the moment you see them coming close to fruition, as this ensures you are prepared to monitor that plant and harvest at the right moment. You can keep track of this information in your plant journal, and keep track of information pertaining to when you specifically harvested a plant, and why. Document whether or not that plant was perfectly ripe when it came time to prepare it, as this will help you become even better at gauging when to harvest your plants in the future. In some cases, you may accidentally pick out underripe or overripe food, due to inexperience. Generally, underripe foods can be ripened in a windowsill, while overripe foods will need to be quickly processed and cooked to avoid them ripening any further. Slightly overripe foods may be able to be moved to the fridge to slow them down, giving you more time to cook with them.

Signs of Readiness for Common Plants

The following 23 plants are commonly grown in backyard gardens. I have included signs of readiness for you to help you get an idea of what to look for if you grow these plants in your garden. This way, you know exactly when to harvest them for optimal nutrition and freshness.

Asparagus

Asparagus is ready to harvest when it is about 6-8 inches tall and as thick as your pinky finger. To harvest asparagus, snap it off at ground level. New asparagus spears will grow from the remaining stocks. 4-6 weeks after your initial harvest, stop harvesting and let the asparagus go wild. This allows the plant to produce food and nutrients for itself.

Beans (Snap)

Before seeds begin bulging out of the pods, snap beans are ready. Check one bean each day, as they should easily snap in two. If they are soft and fail to snap, they are too young. If they are left too long, they will become tough and inedible.

Beets

When you thin out the rows of your beets, the green tops can be harvested and eaten as a form of microgreens. Otherwise, wait until you see the tops of your beets protruding from the soil.

This way, you know they are nice and large, and they are ready to be harvested. You can harvest them sooner, but they will be much smaller.

Broccoli

Broccoli should be picked before the flowerheads bloom. Each of the individual heads should be about the size of a match head, which indicates they are ready. Homegrown broccoli will not be the same size as broccoli from the supermarket, so do not wait for it to grow as large.

Cabbage

When cabbage is gently squeezed, and it feels solid, it is ready to be harvested. Avoid letting your cabbage sit too long, or it will split open and be inedible.

Carrots

Carrots should be harvested when their tops protrude from the soil and appear wide enough to match the variety of carrots you are growing. If you harvest them too soon, you will have small, dwindles of carrots that are not worth eating. Some carrots do better when they are left in the ground for the first light frost, as this is said to help sweeten their flavor.

Cauliflower

Cauliflower should be harvested when it looks full, and the curds of the head still have a smooth appearance. Your homegrown cauliflower is unlikely to grow the size of cauliflower at the supermarket, so look for appearance over size. Leaving cauliflower too long can result in the cauliflower developing brown spots and having a strange, mealy texture when you consume it.

Corn

Corn should be harvested about three weeks after the silks form. The silks will begin to turn dry and brown, indicating it is almost ready. To check for doneness, open one ear of corn and use your fingernail to scrape one of the kernels. If the substance that comes out is milky white, the corn is ready to be harvested.

Cucumber

Cucumbers mature quickly and should be harvested young for optimal freshness. When a cucumber fruit is firm and smooth, it is ready for you to harvest. Waiting too long will result in your fruit becoming bitter, soft, and possibly turning yellow inside.

Eggplant

Eggplants should be firm and shiny when they are harvested, and you should harvest them a few days before you believe they are truly ready, as this is when they have the best flavor. To harvest your eggplant, you should cut it from the stem, rather than pulling it off, as pulling it off can damage the rest of the plant, and possibly the fruit itself, too.

Garlic

Garlic tops begin to fall over and turn brown when the garlic bulbs are ready to be harvested. To harvest garlic properly, dig it out of the ground, don't pull it out of the ground. Brush the dirt off your garlic, don't wash it off, and then dry the garlic before storing it, so it does not turn soft or moldy in storage.

Kale

Kale can be harvested throughout the entire season. To prolong your harvest, pick a few leaves off of each plant at a time, rather than cutting back the entire plant. Leaves are ready when they are deep green and have a firm texture.

Lettuce

Lettuce can be harvested when the entire head feels full and firm. Use a gentle squeeze to pull the head of lettuce out of the soil for eating. Be cautious, as hot weather will cause a head of lettuce to go to seed, which means you can no longer consume it.

Onion

Onions can be harvested when the tops have fallen over. Like with garlic, it is better to dig onions out than to pull them out to avoid damaging them. Dry onions in the sun for a few days before storing them in a cool, dark room in your house.

Peas

Peas are ready to be harvested when the pods look and feel full to the touch—harvest peas before they are fully plumped for a sweeter taste and texture. To know if they are ready, simply try one off the plant and see if it tastes right. If it does, the plant is ready to be harvested.

Potato

Potatoes can be harvested at two different points in the season: "new" potatoes, or young potatoes, are harvested early on and have a softer skin, which makes them excellent for mashed potato dishes. Full-sized potatoes are harvested later, when they have naturally developed a thicker skin, and they do better in long-term storage. When the tops of potato plants start to flower, harvest them for new potatoes. When the tops dry and turn brown and begin to die off, harvest them for full-sized potatoes. Harvest your potatoes by carefully digging the outside perimeter of the plant, moving slowly to avoid damaging the potatoes with your shovel. Once you get close to the plant, use your hands to avoid slicing into any fresh potatoes.

Pumpkin

Pumpkins are ready to harvest when they turn the right color for their variety, and the vines are starting to die off. Use your fingernail to poke the skin of a pumpkin to ensure that it has hardened. If it has not, do not pick it yet. If it has, pick it and store it for up to 1 year in a cool, dark spot.

Radish

Radishes are ready to harvest when the tops of the bulbs begin to protrude above the soil. Regularly harvest your radishes, as they will quickly become tough otherwise.

Swiss Chard

Swiss chard is ready to harvest when the outer leaves look the appropriate color for the variety you have grown, and they are firm to the touch. Only harvest the outer leaves, leaving the center to continue growing. This will elongate your harvest and get you the most swiss chard possible.

Spinach

Spinach can be harvested as individual leaves when the plant is 6 inches tall. Avoid waiting much longer, as they will go to seed quickly, meaning they can no longer be consumed. To harvest spinach, use scissors to cut leaves off at the soil line.

Squash

Summer squash is ready when the skin is still soft, but the fruit is large and firm. Pick them young and often, as they will grow incredibly fast and rot even faster if you are not careful. Winter squash is ready to harvest when they are the appropriate color for their variety. Do not let winter squash be exposed to any frost, or it will damage the squash—harvest squash by cutting it off of the vine.

Tomato

Tomatoes are ready to harvest when they reach the appropriate color and size for their variety. They should still be slightly soft to the touch and should have a sweet tomato scent when you pick them off of the plant. Remove tomatoes from the vine by gently twisting them away from the plant. Store them on the counter for optimal freshness, or preserve them quickly to maintain their fresh flavor.

Watermelon

Watermelon is ready to be harvested when the bottom of the watermelon itself turns a deep yellow color. As watermelons grow, the portion touching the ground will be white until this point. You can also tap a watermelon to listen to the sound it makes, as a watermelon that is ready to be eaten will sound hollow inside. However, not everyone will be able to hear this sound as clearly as others. As soon as the watermelon reaches that deep yellow color, cut it off of the vine and bring it indoors. Watermelons can be stored in a cool, dry place for several weeks, and even months before they are ready to be consumed.

Choosing the Right Time to Harvest Your Plants

Harvesting your produce at the right time of day is essential to the nutritional profile, flavor, and quality of the fruits or vegetables that you are harvesting. Harvesting your produce too late in the day, after it has been hot or after they have been watered, can completely alter the flavor profile of the food itself.

The best time to harvest plants is first thing in the morning before you water them. This is when your plants taste their best, have a crispier and juicer texture, and have their greatest levels of nutrition in them. The science behind this is that the plant has not recently been watered and, therefore, has less moisture in it. This means all of the organic matter that makes up the vegetable itself is concentrated, improving the flavor and texture. If you have recently watered a plant, or if its moisture content is high from humidity, the plant will have more water content in it, which dilutes the flavor and bloats the vegetable.

Removing Produce From the Plant

Removing your produce from a plant largely depends on what type of plant you are harvesting. There are typically three groups plants can be categorized into when they bear food: those which grow food under the ground, those which grow edible leaves, and those which grow food on their branches.

Foods that grow under the ground include potatoes, onions, garlic, radishes, beets, carrots, and any other root vegetable or bulb that humans typically harvest to consume. Because you are consuming that which grows underground, you must be gentle with these plants to avoid accidentally damaging the produce. Careless digging, yanking plants up by their tops, and even walking too close to them can damage these plants. You need to keep your feet well away from the growing area and carefully use your shovel to scoop the dirt away from crawlers like potatoes and beets. For plants that tend to stay in one spot, like carrots, onions, and garlic, you can use your hands to dig up each plant carefully. Proper care at this point will prevent damage in the produce. Any produce that is damaged during harvest should be left to dry with everything else and consumed first unless it is too badly damaged, in which case you can toss it in the compost bin.

Foods that grow as leafy vegetables, such as lettuce, swiss chard, spinach, kale, or any leafy herbs, should all be harvested on a leaf by leaf basis. The one exception is herbs, which can be harvested one small branch at a time, where each branch contains many leaves. Harvesting leafy vegetables and herbs this way ensures that you do not deplete the plant, which means it will continue to produce more leaves throughout the growing season. You can harvest the entire plant itself at the end of the growing season before it goes to seed or dies off. To do this, simply squeeze the plant at the roots and carefully wiggle it up, bringing up the roots and everything. Inside, you can either chop the roots off or place them in a shallow dish of filtered water to keep the plant fresh until you are ready to use it. The one leafy vegetable you should harvest whole every single time is cabbage, as cabbage would not harvest well on a leaf-by-leaf basis.

Foods that grow to hang off of branches or on the ends of vines are harvested based on their size and how they are attached to the plant. Foods like squash, cucumber, eggplants, and watermelon are attached to the plant quite deeply, so you need to cut them off of the plant. After a day or two, the stem will dry out, and you can twist it off of the produce. Foods like tomatoes, berries, lemons, apples, and other similar hanging fruits are able to easily be twisted right off the plant because their attachment is more shallow. They easily twist off without damaging the produce, and this prevents damage to the plant itself, too.

Cleaning Your Produce for Immediate Use

Most produce does not have to be cleaned off unless you are about to consume it. Using a gloved hand or a dry rag and wiping the produce clean is sufficient for storage in a root cellar or in a similar, whole-preserved fashion. When you are ready to cook with your food, rinsing it under cool water is sufficient. Since you did not spray your vegetables with any harsh chemicals, nor did you coat them in anything to preserve their shelf-stability, there is no reason to use harsher cleaners like diluted vinegar or lemon juice. Water is plenty.

Preserving Your Produce for Long Term Storage

Throughout the harvest season, if you are lucky, you are likely to have large amounts of fresh produce. By the end of the season, or possibly even before that, you might find that you have far more than you know what to do with. Having an abundance from your garden is a great blessing, but it can be overwhelming to manage. Fortunately, you do not have to waste your harvest. Plus, it can be much easier to preserve it than you might anticipate. In my other book, *Survival 101: Food Storage*, I discuss several ways that you can preserve your harvest.

Some of the many things you can do to preserve your harvest include cooking it into various recipes and canning your produce, freezing it, dehydrating it, or storing it in sugar, honey, oil, or ash. All of these preservation methods will ensure that you have plenty of produce to get you through the cooler months when your garden does not grow, effectively increasing your self-reliance.

Harvesting Seeds From Your Plants

It is likely that your original seed packets will last quite a while, as most are filled with enough seeds to last at least 2-3 years in a backyard garden. Over time, however, you might find that you are starting to run low and that you need more seeds. While you could purchase more seeds, it is worth noting that this would cost you more money. A great way to save money is to harvest seeds right out of your own garden.

Harvesting seeds from your garden is not nearly as challenging as you might think it would be. For some plants, especially fruits, harvesting seeds is fairly simple: you open the fruit, take a few seeds out, and keep them. For others, getting seeds from your plants requires a bit more effort. Below, I will include information on how you can harvest seeds from some of the most common plants found in a backyard garden.

Lettuce, spinach, kale, swiss chard, and other leafy greens will bolt or go to seed on their own. When this happens, small flowers appear on the plant, which bears the seeds. If you were to pick just two or three of the fuzzy seed heads, you could end up with dozens of seeds to plant in following years.

Tomato seeds can be saved by keeping a healthy-looking tomato and letting it overripen. In the process, the inside of the tomato ferments and destroys any bacterial or viral diseases which could penetrate the seeds and render them useless. This process also breaks down the gel that coats the seeds, making them easier to clean and store. Anytime you harvest seeds from a wet plant such as tomatoes, cucumbers, zucchinis, or otherwise, you should wash them in cool filtered water for a few minutes to remove any plant residue. Then, lay them on a paper towel to completely dry before placing them in a labeled bag and storing them for the year.

To harvest seeds from beans, you want to leave a few beans on the stalk and let the beanstalk dry down. During this process, the plant will lose everything except the pods. If you wait long enough, you will grab the pod off the plant, and the beans would rattle inside because they were already dried out. Pull a dry pod open and try to dent the seeds with your fingernail. If you can't, they are ready to be harvested and stored for future use.

Peppers can be harvested by removing the seed mass from inside the pepper itself and rubbing it to separate the seeds from the mass. One single fruit should supply you with plenty of seeds for the following year.

Broccoli seeds can be harvested by waiting for the plant to produce seed pods. These seedpods are narrow pods that grow out of the plant, and that eventually grow dry and brittle. When they reach this point, they can be harvested from the plant. Break the pod open and carefully extract the seeds, clean them, and store them for future use.

Corn can be kept on the husk and dried through the winter, before being twisted off in the spring to be planted. Corn is a major cross pollinator, so it is ideal to harvest your seeded variety at the end of the season. You should grow as many plants as possible from your retained husks to ensure that you get a strong variety of corn each season.

Cucumbers should be allowed to ferment for a few days so that removing the seeds is easy. This way, the pulp around the seeds releases the seeds, and they can be plucked right out of the fruit itself. Place all cucumber seeds in cool, filtered water and watch for which ones sink and which ones float. Pour off the ones that float, as the ones that sink are higher quality and more likely to grow in following years. Stir the remaining seeds to wash them, then strain and dry them before storing them in an airtight container for the following year.

Squash and pumpkins can have their seeds harvested from directly inside of the squash. Be sure to wash and dry the seeds properly before storing them so that they do not develop any mold or nasty bacteria while they are waiting to be planted.

Beets can be regrown by pulling them out in the fall before the first frost and bringing them indoors. You should only need about three beets to grow a sufficient supply for the following year. Cut their tops off one inch above the crown of the beet and then store them with even access to moisture to prevent rotting. You can layer beets in a box between fresh sawdust or dampened sand to keep them moist. They should be kept between 40-50F. The following year, thin the beets and plant them about two feet apart, keeping the crowns of the beets even with the surface of the soil. They will produce more beets for you this way.

Potatoes can be grown by keeping one or two potatoes in relatively humid conditions and allowing it to grow its own stalks. Once the stalks off the potato are about 6-8 inches long, you can slice the potato so that you have 2-4 different pieces of potato with thriving stalks growing off of them. Plant the potatoes 2 feet apart, with the slice of potato about 4-6 inches underground. Let them grow until the bushy tops are 2 feet tall, or until they start to die off, before harvesting them.

Garlic bulbs should be replanted in the ground in October, and harvested in late July. In July, harvest 1-2 bulbs of garlic with the intention of replanting them in October. Dry the garlic in a shaded spot in your backyard by hanging it and letting it air dry. Do not dry garlic in the sun, as it will cook. Once the garlic is dry, bring it indoors. Keep the bulbs whole for garlic that you intend to replant. In October, break the bulb down into individual cloves and plant each clove 18" apart. New bulbs will grow from them.

Anytime you harvest actual seeds off of a plant, it is always important to ensure they are clean and completely dry before placing them in a bag to be stored. Any moisture retained in the seeds can lead to the seeds inside of the bag molding, which means they are no longer able to be used. Drying seeds can happen on a warm window sill for a few days. All seeds should be laid out on a piece of paper towel in a single layer and left to dry for as long as needed. Once you are confident that they are dry, store them in airtight containers until the following year. Some harvested seeds will last several years in storage; however, it is best to use your seeds within 1-2 years to ensure that they are still fresh. For optimal freshness and improved harvests, store fresh seeds every single year out of your existing crop, and use up all of your seeds the following year.

Chapter 10

Preparing Your Garden for Next Year

Winterizing your garden is a great way to ensure that growing conditions remain optimal in your garden throughout the winter. Winterizing your garden does not take much time, but it will save you plenty of work in the spring and will improve your harvest each year. It is easy to get overwhelmed by the idea of winterizing your garden, especially when you realize there is so much to be done, but the process is not as hard as it seems. Really, the list is longer than it is challenging, and you can easily get it all done within an afternoon or two.

Know When to Winterize

You want to harvest as much as you can from your plants for as long as you can, so there is no reason to rush into winterizing your garden. Wait until the first hard freeze in the fall, then begin winterizing your garden. At this point, any remaining vegetable plants will die off, making it easy for you to transition into winterizing. If it seems like your plants are going to try to hold on a bit longer, don't try to hold on with them. Waiting too long could result in consistent freezing or snowfall, which would impair your ability to winterize your garden properly. This would leave you with a lot of work left to do in the spring.

After your first hard freeze in the fall, you will need to complete several tasks to winterize your garden. Those tasks include: removing dead plants, weeding the garden, destroying diseased plant material, bringing tender plants indoors, turning the soil, and adding fresh compost. I will explain how to do each step below.

Remove Dead Plants

The first and possibly most obvious part of winterizing the garden is removing dead plants. Dead plant matter can be added to your compost, though you should be careful not to overwhelm your compost with too much dead plant matter. If you have more plants than you do space in the compost, remove the leaves off of the plants and add those to your compost and throw the rest of the plant into a brown bag for your local waste pickup. Most regions have a specific compost pick up where they will take your plant matter and add it to the city compost, so the plant matter does not go to waste.

When removing dead plant matter, ensure that you remove everything right down to the roots. Dead roots in the garden take up a large amount of space under the soil, and they make it challenging for next year's plants to find room to grow. Everything from the current year's plants should be completely removed and taken away from the garden.

Weed the Garden

As you remove dead plant matter, remove any weeds that have come through, too. Weeds tend to be rampant in late fall, which makes this a great time to remove all of them fully. Do not leave any weeds behind because if they get turned back into the soil, they will grow even more rampant the following year.

If you have nearby flower gardens, ensure that you cut them back before they go to seed. Some plants, like black-eyed Susan's or chrysanthemums, produce large amounts of seeds, and these can end up in your garden and turn into weeds the following year. Cutting those plants back before this happens ensures that their seeds do not become weeds the following year.

Destroy All Diseased Plant Material

In the fall, there can be an abundance of diseased plant material, even if you had healthy plants all year long. Diseases can often strike dying plants, as they are not as strong or hardy as they were during the lively growing season. As well, dead plants can develop mold and other diseases before they fully dry out. Any plant that has been infected with a disease, even as basic as mold, should be destroyed. Diseased plants that are left to sit in your yard, or that are added to your compost will spread the disease and infect next year's crop.

The best way to destroy diseased plant material is either by burning it or by throwing it directly into the garden. Take care not to let the plant spread or touch anything on its way out, as this can cause it to leave behind spores before you are able to remove it from your garden completely.

Bring Tender Plants Indoors

Some plants, such as herbs and root plants like ginger, can be brought indoors for the winter. These plants are not necessarily ones that die off in the winter, especially if they are kept in proper conditions. By keeping them indoors, you enable yourself to continue harvesting them all throughout the winter. Be sure to harden off tender plants before sending them back outdoors in the spring.

Turn the Soil

After everything is removed from your garden, you want to spend a few minutes turning the soil throughout your entire garden beds. Using your shovel or a hoe, turn the soil as deep as you possibly can. This will aerate the soil and loosen it up so that it is light and fluffy in the spring. You will want to turn it again in the spring, but this ensures maximum aeration for the best growth the following year. After you turn it, refrain from walking over it, patting it, or otherwise compacting it in any way, as you want it to remain as loose as possible.

Add Fresh Compost

Once your soil has been completely turned, you want to start adding fresh compost. Ensure that your compost is completely finished, then add 1-2 inches over your garden bed. Mix this in with the topsoil in your garden, then add another 2 inches of compost to your garden and do not mix it in. This way, as the rain and frost hits and the ground, freezes and thaws several times through the fall and spring, the nutrients from the compost are deposited throughout the remaining soil, and the soil is ready to receive more plants in the spring. This will give you the best, richest soil for future harvests.

Chapter 11

Plant Profiles

There are many plants that are common in home gardens. There are also many plants that are regularly wildcrafted, or picked from the wild, which are edible and safe to use in a variety of manners. Having a clear understanding of what each plant is, where and when it grows, and how it grows is an important part of taking proper care of your plants. It is also a great way to ensure that if you come across wild edibles, you know exactly what they are, and you can safely consume them.

Backyard Garden Plant Profiles

Backyard garden plants are excellent for growing, harvesting, and using to nourish your family. With backyard gardens, you know exactly what you are getting, and you know, for certain, the plants you are consuming are safe and edible. Backyard gardens can include vegetables, fruits, and even flowers, which can be both edible, and used as natural pest deterrent or pollinators to improve the quality of your edible garden. Here, you will discover profiles on the most popular vegetables, fruits, herbs, and flowers to be grown in a backyard garden.

Apples

Apples are a popular backyard crop that grows on large, bushy trees. Many varieties of apples exist, ranging from sweet McIntosh apples to sour crabapples.

Family: Rosaceae

Growing Season: August to October

Zone: 3-5 for hardy plants, 5-8 for long-season apples

Spacing: 30-35 feet

Seed to Harvest: 6-10 years from seedling to bearing fruit

Indoor Seed Starting: Start with cold stratification, then plant the seed ½ inch deep in starting soil.

Earliest Outdoor Planting: Plant seeds outdoors in the early fall, ½ inch deep.

Starting: Moisten a paper towel and lay apple seeds out inside of the paper towel. Place the paper towel in a plastic bag and place the bag in the back of the fridge for six weeks. Check weekly to ensure the paper towel remains moist, mist it if it begins to dry out. After six weeks, some seeds will start, others will not. Plant the started seeds ½ inch deep in warm soil and keep the soil moist.

Growing: Grow seedlings in pots until they are large enough to withstand outdoor elements. Then, when the overnight temperatures are consistently above 50F, plant your seedlings in the ground. Protect them using metal shields or fencing, and continue to water them and grow them until they are large enough to bear fruit. This should take 6-8 years.

Harvesting: When the plant is old enough to bear fruit, it will begin to have fresh fruit available from August to October each year. Simply tug the apples off the tree to harvest.

Problems: Apples are prone to apple scab, fire blight, cork spot, powdery mildew, rust, black rot, and crown rot.

Basil

Basil is a culinary herb that is frequently used in pasta dishes, as well as other hearty dishes. It can be grown in a pot or in your garden beds.

Family: Lamiaceae.

Growing Season: Year-round indoors, mid-April to first frost outdoors.

Zone: 5-10

Spacing: 10-12 inches apart for smaller varieties, 16-24 inches for larger varieties.

Seed to Harvest: 50-60 days.

Indoor Seed Starting: Plant basil ¼ inch deep in moist soil. Start your seeds indoors six weeks before the last frost.

Earliest Outdoor Planting: Plant basil seeds outdoors when the soil is 70F, and nighttime temperatures no longer drop below 50F. Basil thrives in warmer climates.

Starting: Plant seeds directly into moistened soil at ¼ inch deep. Keep the soil moist. Basil should start to surface within one to two weeks.

Growing: Keep plants in a pot indoors in a south-facing window for indoor container growing—transplant seedlings into the garden when proper outdoor conditions are met to grow basil outdoors. After seedlings produce six leaves, prune it back to stimulate branching. Do this every time a new branch develops six to eight new leaves.

Harvesting: Harvest basil before frost comes to avoid losing it to the cold. Use pruning clippings as harvest.

Problems: Fusarium wild, downy mildew.

Beets

Beets are a delicious root vegetable. These juicy vegetables grow in a variety of different colors and tastes, though the most common is an earthy red color, which is used in a variety of dishes.

Family: Amaranthaceae

Growing Season: Late April to late September

Zone: 2-11

Spacing: 2-3 inches apart

Seed to Harvest: 45-65 days

Indoor Seed Starting: Not recommended unless starting in their permanent containers.

Earliest Outdoor Planting: Late April, after the ground has warmed up to 70F+.

Starting: Plant beet seeds directly into the soil at ½ inch to 1 inch deep, 2-3 inches apart. Take care not to crush the seeds because they are tender. Gently cover the seeds with a light layer of topsoil and water well. Keep seeds watered with about 1 inch of water per week, and be patient. They take a while to germinate. 2-3 weeks after they sprout, thin them out to 1 plant every 2-3 inches. You can eat the thinned plants in a salad or on a sandwich.

Growing: Water maturing beets with 1 inch of water per week. Otherwise, let them be. When beet shoulders start to poke above the soil, they are ready to be harvested.

Harvesting: When beets are about 3 inches in diameter, use your hands to dig them out of the soil carefully. Do not wash them until you are ready to use them. Cut tops off 1 inch away from the crown of the beet.

Problems: Aphids, dampening off, flea beetle, leafhoppers, and seed rot.

Broccoli

Broccoli is a delicious plant that grows on stalks, with small bulb-like flowers that sprout on the ends. You need to harvest it before these flowers bloom; otherwise, they are no longer edible.

Family: Brassicaceae

Growing Season: Late April to mid-October

Zone: 3-10

Spacing: 3 inches apart as seedlings, 12-20 inches apart as maturing plants

Seed to Harvest: 100-150 days

Indoor Seed Starting: Start seeds indoors 6-8 weeks before the last frost by planting them ½ inch deep and 3 inches apart in containers, or in 3-inch starter pots. Transplant seedlings at 4-6 weeks old, when they are tall enough to be thinned out at the same time.

Earliest Outdoor Planting: In late April.

Starting: Start seedlings by planting them ½ inch deep in starting soil. Keep seeds 3 inches apart. Once the plant reaches about 2-3 inches tall, thin them out, so they are 12-20 inches apart, depending on the size of your chosen variety. When transplanting your seedlings, place them in an area where they will get 6-8 hours of sunlight per day.

Growing: Fertilize broccoli three weeks after planting it into the garden. Keep the soil moist using at least 1 to 1 ½ inches of water per week, but increase that amount if you are experiencing a drought. Avoid getting water on developing broccoli heads to discourage rotting. Use mulch to keep soil moist and prevent weeds, as broccoli roots are shallow, and weeds will quickly kill them off.

Harvesting: When the plant is just about to go into flower, cut the heads off of the plant, taking at least 6 inches of the stem with it. Broccoli will develop small yellow flowers if it is left too long, and, if it does, it will taste awful. If you see any flowers develop, harvest the plants immediately.

Problems: Aphids, cabbage loopers, cabbage root maggots, cabbage worms, clubroot, downy mildew, white rust, whiteflies, and nitrogen deficiency.

Carrots

Carrots are a leggy root vegetable that comes in many colors. They liven up a plate with their appearance and can be eaten raw or used in soups, stews, steamed dishes, salads, or anything else where vegetables might be enjoyed.

Family: Umbellifers

Growing Season: Early spring to mid-November.

Zone: 3-10

Spacing: 3-4 inches between plants, 1 foot between rows

Seed to Harvest: 70-80 days

Indoor Seed Starting: Not recommended.

Earliest Outdoor Planting: Plant seedlings outdoors 3-5 weeks before the last spring frost.

Starting: Sow seeds ¼ inch deep, keeping the seeds 3-4 inches apart. Ensure the rows are 1 foot apart. Cover seeds with vermiculite or fine compost to ensure the surface remains soft and easy to sprout through. Sow new seeds every three weeks for multiple harvests.

Growing: Mulch carrots to retain moisture using a small amount of a light mulch. When seeds are 1 inch tall, thin them to 3-4 inches apart by snipping unwanted plants with scissors, avoid pulling them out, as this will damage the roots of the remaining carrots. Water one inch per week until the roots begin to mature, then increase your watering to two inches per week.

Harvesting: When the carrots are your desired size, harvest them. Keep in mind that smaller carrots generally taste better, and those that have undergone one light frost will taste sweeter.

Problems: Aster Yellow Disease, Black canker, carrot rust flies, flea beetles, root-knot nematodes, and wireworms.

Chives

Chives are a herb that is commonly grown to complement dishes in a way similar to green onions. Chives tend to have a more tangy or onion-like flavor to them, making them a unique addition to many modern dishes.

Family: Amaryllidaceae

Growing Season: February to September

Zone: 3-9

Spacing: 2 inches between plants

Seed to Harvest: 60 days

Indoor Seed Starting: Start seeds in a container indoors in February.

Earliest Outdoor Planting: Plant seeds in a container outdoors in April.

Starting: Plant seeds no more than ¼ inch deep and keep about 2 inches between plants. Cover with a very thin layer of soil. Thin seedlings to 4-6 inches apart once they emerge.

Growing: Ensure consistent watering to maintain a consistent harvest. However, chives are drought-tolerant, so they can dry out between watering. Drying the chives out between watering will help increase their flavor potency.

Harvesting: Snip a few chive stalks off the plant 1" above the base as needed. Bring the pot indoors for year-round harvesting.

Problems: Bulb rots, fungal leaf spots, mildew, onion fly, rust, smut, thrips, and white rot.

Green Beans

There are many different beans that you can grow in your garden. Most beans have similar growing requirements and needs. This particular profile is specifically for green beans.

Family: Fabaceae

Growing Season: early spring to first frost

Zone: 3-10

Spacing: 2 inches between plants, 18 inches between rows

Seed to Harvest: 50-60 days

Indoor Seed Starting: Not recommended. Fragile roots damage during transplanting, killing off the plants.

Earliest Outdoor Planting: Any time after the last frost, when soil is at least 48F. Avoid planting too early, as any cold snap will kill green beans.

Starting: Plant seeds 1 inch deep and 2 inches apart, in rows that are 18 inches apart. For sandy soil, plant them about 1.5 inches deep. Place seeds directly into the ground, lightly covering them in topsoil. For pole bean varieties, set up trellises or caves prior to planting to avoid damaging the delicate roots of a mature plant. Sow new beans every two weeks for continuous harvesting all season long.

Growing: Place mulch around the beans to retain moisture, but keep the mulch well-drained. Give them about 2 inches of water per week, with regular watering. Water on sunny days, only, to prevent the foliage from remaining wet and picking up diseases.

Harvesting: Harvest beans first thing in the morning by picking them off the bean plant. Pick fresh beans daily to encourage the plant to continue growing more. Be careful not to tear the plant when removing beans from it. Pick before the beans are bulging inside.

Problems: Alternaria leaf spot, anthracnose, bean rust, black root rot, fusarium root rot, white mold, bacterial blight, bacterial brown spot, halo blight, damping-off, mosaic.

Kale

Kale is a leafy green that is known for being one of the most nutritious varieties of leafy greens available. Although it has a somewhat tough texture, it can be cooked into larger dishes to make it more palatable.

Family: Brassica oleracea

Growing Season: March to first hard frost

Zone: Winter hardy to zone 6

Spacing: 8-12 inches

Seed to Harvest: 55-75 days

Indoor Seed Starting: Not necessary.

Earliest Outdoor Planting: Direct-plant seeds outdoors 3-5 weeks before the last frost, as long as temperatures are not dipping below 20F. If they dip below 20F, cover your rows of seeds with proper covers to protect them from the cool nights.

Starting: Plant seeds directly into well-drained, light soil at ¼ to ½ inch deep. Place 3-4 seeds per spot, and thin the seeds around two weeks old to proper spacing. Water seeds, so the soil is moist but not drenched.

Growing: Keep kale well-watered with 1 to 1.5 inches of water over the plant every single week. Continuous-release plant food can be used to improve kale's growth. Mulch the soil after the first hard freeze in the fall to elongate growth. Some plants will produce into the winter if mulched.

Harvesting: When leaves are the size of your hand, clip the outside leaves off of the plant and leave the rest alone. This will encourage the plant to continue producing. At the end of the season, clip the entire plant back.

Problems: Alternaria leaf spot, damping-off, downy mildew, black rot.

Lettuce

Lettuce is rich with vitamin A. It is one of the easiest garden vegetables to grow and can be harvested throughout the entire growing season for fresh salads, sandwiches, or other culinary dishes.

Family: Asteraceae

Growing Season: Early April to Early September

Zone: 4-9

Spacing: 12-15 inches between rows. Leaf lettuce: 4 inches between plants, loose-headed lettuce: 8 inches between plants, firm-headed lettuce: 16 inches between plants.

Seed to Harvest: 30-60 days, depending on the variety.

Indoor Seed Starting: Start seedlings indoors 4-6 weeks before last spring frost for an early crop.

Earliest Outdoor Planting: Direct sow seeds when the soil remains above 40F in the spring.

Starting: Plant seeds ¼ inch to ½ inch deep. They should germinate between 7-10 days. After two weeks, thin seeds out. When seeds have 4-6 mature leaves each, and their root system is developed, transplant them to their outdoor location. Do not transplant before the last frost—water thoroughly after transplanting them. Start a small number at a time, and start new seeds every other week for a staggered harvest to avoid having too much lettuce.

Growing: Fertilize after three weeks of being planted outdoors. Keep soil moist but well-drained. When lettuce begins looking wilted, sprinkle them with water, even if it is in the middle of a hot day. Organic mulch can be used to keep the base of a lettuce plant moist, which will help it thrive.

Harvesting: Just before maturity, remove the outer leaves, dig up the plant, or cut it back one inch above the soil surface. Leaving the center or the 1 inch base of the plant will allow for more lettuce to grow from that plant.

Problems: Anthracnose, leaf drop, powdery mildew, septoria leaf spot, big vein.

Mint

Mint is a hardy herb that will multiply rapidly, so it is best grown in a container and contained to prevent it from taking over your garden. This plant can turn into a weed extremely fast.

Family: Lamiaceae

Growing Season: Early spring to late fall; year-round in warm climates

Zone: 3-10

Spacing: 2 feet

Seed to Harvest: 90 days

Indoor Seed Starting: Plant a few seeds in the center of a 2-foot pot. Keep the soil moist but well-drained.

Earliest Outdoor Planting: Plant outdoors after the last frost, in a 2-foot container. Keep the soil moist but well-drained. Winterize the pot or bring it indoors during cold months to prevent freezing.

Starting: Plant a mint seed ¼ inch to ½ inch deep in its container. Keep it moist, but not saturated. Seeds should start to sprout within 5-10 days.

Growing: Water daily, as mint likes to remain moist. It grows naturally near streams, meaning it does not like dry soil. Keep it in a mostly-sunny location or in a location with indirect light depending on the variety you are growing.

Harvesting: Harvest frequently to keep the plant healthy and thriving. Harvest the entire mint plant at once just before it starts flowering, cutting it back to 1 inch above the ground. The plant can be harvested 2-3 times per year, as long as growing conditions are right.

Problems: Anthracnose, leaf spot, powdery mildew, rust, stem canker.

Oregano

Oregano is a popular culinary herb that is frequently used in pasta dishes, on pizzas, and in other common dishes. It is well-liked in the kitchen, and not hard to grow yourself.

Family: Lamiaceae

Growing Season: Late-spring to mid-summer

Zone: 4-10

Spacing: 8-10 inches apart

Seed to Harvest: 80-90 days

Indoor Seed Starting: Plant indoors 6-10 weeks before the last frost in a warm, south-facing window. Use seeds or cuttings to start your plants.

Earliest Outdoor Planting: Plant seeds directly outdoors after the last spring frost, when the soil is 70F or higher.

Starting: Plant seeds directly in the garden. Give fresh seeds a good watering, then let the soil dry out between watering—thin seeds out at two weeks, giving 8-10 inches of space between maturing plants.

Growing: Regularly trim the bush to encourage more growth. Let the soil dry between watering for optimal growth. Let the plant go to seed and naturally grow back on its own between growing seasons for easy growth.

Harvesting: Harvest leaves as they are needed. Harvest the entire plant mid-summer before it goes to flower, freezing extra leaves for fresh use at a later date. You can also dry the leaves and store them in an airtight container.

Problems: aphids, root and stem rot, spider mites.

Parsley

Parsley grows light, feather-like leaves and is frequently used in soups, salads, and sauces, as well as for garnish on fancy dishes. It is also rich in iron, vitamin A, and vitamin C.

Family: Apiaceae

Growing Season: Mid-February to first hard frost

Zone: 4-9

Spacing: 6-8 inches between plants

Seed to Harvest: 70-90 days

Indoor Seed Starting: Start seeds indoors in mid-February or mid-March.

Earliest Outdoor Planting: Mid-May, when the soil is 70F or higher.

Starting: Parsley is slow to start, so the sooner you start, the larger your harvest will be. Plant seeds in moist, rich starting soil about ¼ inch to ½ inch deep. Keep seedlings 6-10 inches apart from the start, or plant in their own pods. Keep the area weed-free. Parsley will begin to sprout around three weeks after it has been planted. Indoors, use a fluorescent light to aid growth, but keep the light at least 2 inches above the tallest leaves at all times. Any closer will burn the leaves and damage or possibly kill off your plants.

Growing: Water the seeds often, as they will not germinate if they dry out. After they have grown, evenly water the plants with about 1 inch to 1.5 inches of water per week.

Harvesting: When leaves have separated into three segments, cut the outer portions of the plant off as needed. Put fresh leaf stalked in water in the fridge to keep them longer, or freeze fresh leaves for even longer storage. You can also dry parsley, crumble it, and store it in an airtight container for future use. Bring parsley indoors over the winter if you want to keep it alive year-round.

Problems: Black swallowtail larvae, carrot fly larvae, celery fly larvae, leaf spots, stem rot.

Peaches

Peaches are a juicy stone fruit that can be used in many different recipes. From desserts or drinks to side dishes or toppings, peaches are incredibly versatile.

Family: Rosaceae

Growing Season: Late-June to mid-September

Zone: 4-9

Spacing: 15-20 feet

Seed to Harvest: 3-4 years

Indoor Seed Starting: Clean and soak the peach pit. Pour room temperature water into a small bowl and soak the peach pit for 30 minutes. Do not open the peach pit to remove the seeds from inside, leave it intact. Remove the peach pit from the room temperature water, loosely wrap it in moist sphagnum moss, and seal the entire thing in a zip-top bag. Refrigerate the bag between 33-40F for eight weeks without disturbing it. This is called cold-stratification. Remove the bag from the fridge after eight weeks and let the stone rest. Moisten potting soil with clean water until it feels like a damp sponge to the touch. Fill a 4-inch starting pot with the moistened soil and plant the seed 1 inch into the soil, then lightly cover it. Keep the pot in a warm area with 70F temperature and indirect or filtered sunlight. When the soil begins to feel dry, lightly moisten it to aid germination. Seed should begin to sprout within 1-3 weeks.

Earliest Outdoor Planting: In late fall before the ground is too hard, plant the peach pit 3-4 inches deep in the spot where you want the tree to grow. Cover it lightly with topsoil, then cover the entire site with straw or similar mulch. Water it, and then only water it again anytime the site is completely dry, until the winter hits. During winter, leave it alone. When the spring comes, resume watering the pit whenever the soil starts to dry out. The seed should begin to sprout in early spring.

Starting: Starting seeds indoors using the aforementioned conditions gives you the best chance to water over the seed and help it grow, or remove the seed if it was not viable. Starting seeds outdoors works if you are unconcerned with how the process goes, or if you are okay with starting over again the following year if it doesn't work out.

A viable alternative is to purchase a young 1-year-old seedling from a garden center and plant it. You should plant young trees immediately upon getting them home to avoid shocking the tree any further, so they are more likely to thrive where you place them. Always dig the hole a little bigger than the root system, and never fertilize a peach tree on planting day.

Growing: A mature peach tree will require about 35-40 gallons of water per day to survive. Avoid overwatering, as it can lead to root rot. Fertilize young trees six weeks after planting in a circle around the tree, about 18 inches away from the trunk of the tree itself. In the tree's second year, place ¾ pound of nitrogen fertilizer around the base of the tree in the spring and again in early summer. In the tree's third year, place 1 pound of nitrogen fertilizer around the base in the late spring. Do this each year as the tree matures. Never fertilize a tree within two months of the first fall frost, or while the fruit is maturing. Prune trees into an open center shape each summer, well before the harvest, as this encourages them to thrive. Not pruning trees will cause branches to break and will reduce your harvest.

Harvesting: Peaches will be ready from mid-June to early-September, depending on your tree and your growing zone. Harvest regularly, when the peaches are the right color and ripe for the variety you have grown. This will encourage the tree to produce more peaches, while also ensuring you have the highest quality and freshest fruit.

Problems: Aphids, borers, brown rot, Japanese beetles, leaf curl, leafhoppers, mosaic viruses, powdery mildew.

Spinach

Spinach is a cold-hardy leafy green that can be planted in early spring, fall, and in areas that have more mild winters. It is rich in vitamins A, B, and C, as well as iron and calcium.

Family: Amaranthaceae

Growing Season: Early spring and early fall

Zone: 2-9

Spacing: 1" between seeds.

Seed to Harvest: 42-55 days.

Indoor Seed Starting: Only if you have a cool area for spinach to grow.

Earliest Outdoor Planting: Early February until late May, late September to late November, or late September to late May if you are in a warmer climate.

Starting: Plant seeds directly in your container or garden, with about 12 seeds per one foot of row. Seeds should be sown ½ inch to 1 inch deep, and lightly covered with topsoil. Alternatively, you can sprinkle them over a container and lightly cover the container with a few handfuls of soil. Water the seeds well to get them going.

Growing: Thin seedlings to 3-4 inches apart once they start to grow, otherwise, leave them alone. Spinach roots are shallow and fragile, which means too much tampering with them will result in the crop dying off. Water your seeds regularly.

Harvesting: Harvest as soon as they have small ½ inch or 1-inch leaves for "microgreens" or allow them to grow until the leaves reach your desired size. Avoid waiting too long, as larger leaves are bitter and will not be as enjoyable. Harvest the whole plant at once, or pick leaves off the outside of the plant so inner leaves can develop longer.

Problems: Blight, bolting, downy mildew, leaf miners, and mosaic virus.

Strawberries

Family: Fragaria

Growing Season: June

Zone: 2-10

Spacing: 20 inches

Seed to Harvest: 28-42 days

Indoor Seed Starting: Start in a greenhouse to ensure plants get optimal lighting conditions for growth. Plant seeds in soil ½ inch to 1 inch deep and keep the soil moist. Place it in a south-facing window, so the seeds get at least 6 hours of direct sun per day.

Earliest Outdoor Planting: Plant seedlings outdoors when the roots are no longer than 8 inches.

Starting: Seedlings should be planted outdoors 20 inches apart, with rows 4 feet apart. Strawberries like to sprawl, so you want plenty of room for runners to grow. Ensure holes for plants are deep and wide enough to accommodate for root system without bending or squishing it—water plants well after planting.

Growing: Mulch the strawberry beds to reduce water needs, as well as to prevent weeds from growing. Weed frequently to avoid weeds taking over, as they will compete with and kill off strawberry plants. Keep moisture high at the surface of the soil, as strawberry plants have shallow root systems. Fertilize once when plants begin to grow, but before they have produced berries. In the winter, if your temperature drops below 25F, cut foliage back to 1 inch, mulch the plants 4" deep, and cover with black plastic.

Harvesting: Harvest fruits when they are fully red and ripe every three days. Cut berries by the stem, never pull them off, or you will damage the plant.

Problems: Gray mold, Japanese beetles, powdery mildew, slugs, and spider mites.

Swiss Chard

Family: Amaranthaceae

Growing Season: Mid-April to first hard frost

Zone: 3-10

Spacing: 2-6 inches apart, in rows that are 18 inches apart

Seed to Harvest: 45-55 days

Indoor Seed Starting: Not recommended.

Earliest Outdoor Planting: Mid-April to 40 days before first fall frost

Starting: Start seeds in fresh starting soil, ½ inch to 1 inch deep. When plants reach 3-4 inches tall, thin them out to 6-8 inches apart in the garden, if you have too many, snip back abundant seedlings and consume them as microgreens.

Growing: Water evenly and consistently for abundant growth. Mulch can be used to maintain moisture if you live in a particularly dry climate.

Harvesting: Harvest plants before they reach 1 foot tall, but when they are 6-8 inches, preferably. Harvest outer leaves to let inner leaves have more time to grow. This will extend your harvest season.

Problems: Aphids, cercospora leaf spot, and slugs.

Peas

Peas are a great early crop that offers an excellent nutritional profile, while also being quite tasty. There are three different types of peas you might grow, including English peas, snow peas, and snap peas. All peas require the same growing conditions and care.

Family: Fabaceae

Growing Season: Early spring to first hard frost

Zone: 2-9

Spacing: 2 inches between plants, 12-24 inches between rows.

Seed to Harvest: 60-70 days

Indoor Seed Starting: Not recommended, as peas have fragile root systems that will damage during transplanting.

Earliest Outdoor Planting: 4-6 weeks before last frost, when soil is at least 45F. Wait until snowmelt and spring rains dry up a bit, as drenched soil will render seeds unviable.

Starting: Set a trellis before planting seeds to avoid disrupting seeds or the fragile roots of maturing pea plants later on—plant seeds directly in the ground, 1 inch deep, and with recommended spacing requirements. Water the soil, then use a chopstick or similar sized branch to lightly press seeds back into the soil if any floated to the surface. Peas should germinate within 1-2 weeks.

Growing: Peas will grow between 2-8 feet in height, depending on the variety you are growing. Let your peas dry out a little bit between watering, keeping watering sparse. If plants are wilting, water a little extra, especially in particularly dry weather. You do not want the soil completely dry, but you also do not want it to remain heavily moist at all times, either. Completely dry soil will result in no pea pods, and heavily moist soil will result in root rot. Ensure pea bed remains well weeded, as weeds will consume the sparse amounts of water that the peas need. Avoid damaging fragile pea roots by weeding the garden by hand.

Harvesting: Harvest peas by using one hand to stabilize the vine and the other to lightly tug the pea away from the plant. Peas should be picked first thing in the morning when they are a bright, beautiful green. If you wait too long, pick the peas, dry them, and store them to be re-hydrated for winter soups.

Problems: Aphids, downy mildew, fusarium wilt, Mexican bean beetles, powdery mildew, root-knot nematodes, and wireworms.

Tomatoes

Tomatoes are a versatile plant that can be eaten as is, fried, added to soups or stews, turned into salsas, or cooked down into sauces or pastes. Many varieties of tomatoes exist, each of which offers a different culinary experience. All of which, however, requires relatively similar growing conditions.

Family: Solanaceae

Growing Season: late spring to late summer.

Zone: 4-11

Spacing: 2 feet for small bush tomatoes, 3-4 feet for large bush tomatoes. 4 feet between rows.

Seed to Harvest: 45-55 days

Indoor Seed Starting: Start seeds indoors 6-8 weeks before last frost in 4-inch starting pots. To do so, plant tomato seeds ½ inch deep in moist starting soil and keep them in bright, direct sunlight for at least 6 hours per day. You can use a fluorescent light at least 2 inches above the tallest portion of the plants for 6 hours a day to emulate sunlight if needed.

Earliest Outdoor Planting: Not recommended to start tomato seeds outdoors. Transplant can take place after the last spring frost, when the soil has warmed to around 70F.

Starting: After the last frost, dig one foot into your garden bed and mix in aged manure or completed compost—Harden off seedlings for one week before transplant day. Place tomato cages in the soil before planting your seedlings to avoid damaging tomato roots later on. Apply

2-3 pounds of fertilizer that is *NOT* rich in nitrogen per hundred square feet of garden area to the garden soil. Dig holes for the tomato plants to go into, ensuring the holes are larger than the established root system. Using your index finger and thumb, pinch off the lower branches of seedlings to increase the stem length by the base. Bury up to 2/3 of the plant. Thoroughly water the transplants once they are all placed. If you can, shade them with a darker garden cover for the first week or so for a few hours during the heat of the day to continue hardening them off. This prevents the leaves from becoming too dry, which would cause the plants to die.

Growing: Generously water seedlings, then water them with about 2 inches of water per week. Deep watering enables a stronger root system, which will strengthen the plant. Early morning watering is best to ensure plants remain well-hydrated throughout hot summer days. Use 2-4 inches of mulch to prevent soil from splashing over the lower tomato leaves. Large, flat rocks can also be kept over areas of the soil to help retain moisture, as the rocks prevent the soil from fully drying out. Use hands to pull out weeds to avoid damaging the tomato plants roots.

Harvesting: Let your tomatoes hang on the vine as long as possible. Firm, deep red tomatoes are ready to be picked, regardless of size. Tomatoes that fall off before they are ready should be kept in a paper bag, stem up, in a cool, dark place to finish ripening. Tomatoes kept in the window will rot before they ripen. If the frost comes and your tomatoes are not done ripening, pull the entire plant by the root and hang it upside down either in your basement or in your garage. Tomatoes will ripen this way, and you can pick the remaining fruits. Refrain from placing fresh tomatoes in the fridge, as they will lose their flavor and texture.

Problems: Aphids, blossom-end rot, cracking, flea beetles, late blight, mosaic virus, tomato hornworms, and whiteflies.

Wild Edible Profiles

Wild edibles consist of any wild plants that can be harvested for consumption. Harvesting wild edibles can seem daunting, particularly when you do not know exactly what to look for. However, there are many wild edible plants that have excellent nutritional profiles, delicious flavor, and that are easy to find. Wildcrafted plants also tend to have high nutritional value, and they are relatively easy to find and harvest.

The best way to safely source wild edibles in your area is to pick up a local book on wildcrafting herbs and plant life, or to work with a local herbalist who is well-versed on your local flora and fauna. Going on plant walks with local herbalists or those educated in wildcrafting can help you safely and confidently ID plants and harvest them, without over-harvesting them.

In the wild, many plants have poisonous alternatives that appear similar, while others exist where some of the plant is edible where other parts of the plant are dangerous. Learning this information ahead of time ensures you can feel confident that you are getting the right plant from the forest.

Chapter 12

Resources

Gardening is a vast topic, and it will certainly take more than one guidebook to get you through the process of mastering your gardening skills. Understand that you live in an area that is unique, with growing conditions that are unique to you. Even within your own community, everyone experiences a wide range of growing conditions based on their surroundings. Some gardens will be north-facing rather than south-facing, or some may be situated on the east or west side of the lawn. Other gardens may be shaded by large trees or competing with wild flora or fauna for optimal growing conditions, while some may seem to have the best possible location available.

Learning about your unique gardening space, as well as your zone and your own gardening area, is the best way to ensure that you learn everything you need in order to master gardening in *your* space. Again, keep a plant journal available so you can confidently keep track of everything you learn, either from other gardeners and resources or from trial and error. This way, you have access to this valuable information to help keep you growing a stronger and more vivacious garden every single year.

Below are some great resources to help you get started if you are in the United States of America. If you are not, I encourage you to look for similar resources in your area so you can gain access to support with growing raised bed gardens in your yard.

American Horticultural Society

A resource designed to support both backyard agricultural gardening, as well as different styles of gardening such as landscaping and decorative gardening.

- American Horticultural Society Main Page: https://ahsgardening.org/

Baker Creek Heirloom Seeds

An online seed company that offers rare and unique heirloom seeds for sale. Seeds can be purchased directly online.

- Baker Creek Heirloom Seeds Shop Page: https://www.rareseeds.com/

Burpee

An online seed catalog with a variety of classic and high-quality seeds that are known for offering high yields every time. Excellent for beginner gardeners.

Burpee Main Page: https://www.burpee.com/

Gardener's Supply Company

An American garden supply company that also offers extensive support in developing your backyard garden.

- Gardener's Supply Company Main page: https://www.gardeners.com/

National Gardening Association

An excellent all-around resource for gardening in your backyard in any zone in the United States of America. Offers a membership for additional benefits.

- Garden Research Division: www.gardenresearch.com
- Grants and Awards Division: www.kidsgardening.org
- Educational Resources: www.garden.org

Seeds of Change

An online seed shop with 100% certified organic seeds. No pesticides or GMOs are present in these seeds. Heirloom varieties are available.

- Seeds of Change Main Page: https://www.seedsofchange.com/seeds

Seed Savers Exchange

An online seed shop with rare seeds, as well as a membership that provides access to community exchange boards. Trade seeds with gardeners from around the world using an old-fashioned barter system.

- Seed Savers Exchange Main Page: https://www.seedsavers.org/

Southern Exposure Seed Exchange

A massive seed exchange that is based in Virginia but ships all over the United States. Many seeds are historical varieties such as peanuts, butterbeans, cowpeas, and roselle. They have a heritage collection of seeds for anyone looking for historical favorites.

- Southern Exposure Seed Exchange Main Page: https://www.southernexposure.com/

Territorial Seed Company

An online seed store with seeds optimized for Northwestern states. They also carry tools, books, and other guides to help you grow your backyard garden. All seeds have a full guarantee, so it is safe to try new products through them.

- Territorial Seed Company Main Page: https://territorialseed.com/

USDA National Agricultural Library

An all-around gardening support system designed by the government featuring information provided by national researchers.

- Home Gardening: www.nal.usda.gov/topics/home-gardening
- Community Gardening: https://www.nal.usda.gov/afsic/community-gardening

Conclusion

Congratulations on reading *Survival 101: Raised Bed Gardening!* Gardening is a valuable skill to have, as it allows you to grow sustainable, nourishing food for yourself and your family. When times are tough, gardening can secure your access to healthy, high-quality food that can be consumed fresh or preserved for later consumption. Even when times are not tough, gardening is important as it allows you to reduce your grocery bill. It allows you to maintain a well-stocked pantry for when tough times fall. Since we never know when those tough times may fall upon us, it is always better to be safe rather than sorry.

Most vegetable gardens produce food in as little as a few weeks and will continue producing food throughout the entire growing season. Some plants will even produce until late into the fall or through the winter, depending on where you live. While it may take time, effort, planning, and some learning curves, gardening is always well-worth it.

2020 hit us like a wrecking ball, reminding us that while our modern supply chains are handy, they are not always reliable. When the system goes down, you must know how you are going to sustain yourself in the meantime. While gardening will not put meat in your freezer or give you immediate access to foods, it will give you access to high quality, nourishing foods for long periods of time. In dire circumstances, one can survive off of a vegetarian diet from the yield of their garden alone. At the very least, you know you will be protected until local supply chains open up again.

It is essential to understand that your focus should not just be on growing food for challenging years, but for all years. You truly never know when a devastating situation may strike your family, community, country, or the globe, leading to shortages and challenges, as we have seen this year. In hindsight, no one could have seen this coming, and it has undoubtedly left a tidal wave of damages that will take years to recover from.

If you, like many others, are flocking to gardening and other practical self-sufficiency skills because of the current global situation, I want to take a moment to warmly invite you to this lifestyle. I suspect you will find a great deal of comfort, relief, and support here. It is valuable to remember that, at the end of the day, our survival relies on us exclusively, and we must always know how to sustain ourselves in the event that our modern world falls apart. Even if you are not a prepper person, knowing how to sustain yourself is critical.

If you would like to take your practical self-sufficiency skills a step further to prepare yourself and your family for whatever might come your way, I have three additional books that might interest you. These books were designed, like this one, to provide you with guidance on how you can survive any situation you might happen across. From practical urban environment survival guides to surviving in the woods, you will discover everything you need to navigate any situation you might find yourself in.

These three additional books are:

- *Survival 101: Bushcraft*
- *Survival 101: Beginner's Guide 2021*
- *Survival 101: Food Storage*

No matter what situation you find yourself in, knowing how to garden in raised bed gardens is valuable. This form of gardening can be done on an apartment patio, in the backyard of your urban home, on an acreage, or in the woods if you find yourself having to survive off-grid for any reason, at any point in your life. It is reliable, consistent, and much easier than in-ground gardening, and can provide you with an excellent yield year over year. As you get started, I want to remind you to take your time and be patient with yourself. If you are brand new to gardening, having less diversity but a great yield is better than having a larger diversity of plants but with lower yields. Each plant requires specific care, and it can be challenging to keep up with everything if you are new to the experience. Take your time and build up your knowledge, and before you know it, you will be planting thriving gardens with massive diversity.

Before you go, I want to ask if you could please take a moment to review this book on Amazon Kindle? Your honest feedback would be greatly appreciated!

Thank you, and have fun gardening!

Survival 101 Food Storage:

A Step by Step Beginners Guide on Preserving Food and What to Stockpile While Under Quarantine For 2021

Rory Anderson

Introduction

In 2020 we faced a global pandemic, mass unemployment levels, and destruction in our supply chains. We have endured a number of different challenges already. Unfortunately, it seems like those challenges are going to continue for the foreseeable future. In the past, similar pandemics had lasted between 18-24 months before everything settled, and it often took years after for things to return to normal. If our current state of affairs turns out anything like historical events have in the past, we will have quite a few months, or even years, left to endure everything that has fallen upon us.

One of the biggest things people are worried about, *you* are likely concerned with, is the destruction in our supply chains. Suddenly, there is a considerable level of uncertainty in what will be available and when. Further, people are becoming increasingly more skeptical about where their products come from and are hesitant to trust products that are being imported from overseas suppliers. What does this mean for you? It means you need to take your supply chains into your own hands.

Modern life is convenient, and it has pampered us, that's for sure. Being able to go to the grocery store and get anything you want, whenever you want, and feeling confident that it will never run out is a luxury that many do not have. It is a luxury that those before us didn't have, either. The reality is, our modern way of life may be lavish, but it is certainly not sustainable. As we are currently seeing, just one thing can completely derail the system and leave hundreds of thousands of people without everyday comforts.

When the system falls apart, what are you left to do?

You either sustain yourself, or you die.

It may sound intense, but it's true. If you do not have food, water, shelter, or any other means necessary to live, you will not live. Money may have bought your way to survival in the past, but if the supply chain is damaged and the products are unavailable to purchase, money isn't going to get you very far. Not to mention, the mass unemployment rates mean that most people cannot even afford to sustain themselves as it is.

The solution to the destruction in the supply chains and the mass unemployment rates is to do it yourself. You need to figure out how you can meet your own needs, and continue meeting them for the long haul. This way, you no longer have to worry about a finicky system that seems to be sitting on the brink of ruin. That starts with knowing how to manage your food supply.

Survival 101: Food Storage will describe, in detail, how you can save money and sustain yourself and your family for the foreseeable future through the power of food preservation. You will discover how you can buy inexpensive produce and meat cuts, preserve them safely, and use them to create meals for months and even years to come. In doing this, you will drastically cut back on your food bill while also ensuring that everyone in your family has plenty of food to eat.

Food storage and preservation may be intimidating to some, especially when you consider the many bacteria and parasites that can live in our food sources. Fortunately, with proper storage and preservation methods, you can destroy any chance of your food containing such things, and confidently consume the food you've stored for yourself and your family. Chances are, you will also discover that homemade preserves are far tastier than those that you can buy at the store. You may even, long after all of this is over, continue with this self-sufficient way of eating. I'll let you decide that for yourself when the time comes.

In the meantime, let me show you how you can confidently stock your pantry, fridge, and freezer with enough food to sustain your family through quarantine and beyond.

Chapter 1

Preservation Methods

Preserving food is done in a variety of ways. Think about every time you have walked through the grocery store and saw food on the shelves. Every single item in that store has been preserved in a way that made it shelf-stable for as long as it needed to be in order for it to be sold and then stored in your kitchen until you were ready to eat it. Yes, each item had an expiry date. All food, no matter how well-preserved, will. However, those expiry dates were *lengthened* by the preservation methods used. Some preservation methods result in shorter expiry dates because the supplier knows that those foods will be consumed quickly. Therefore, lengthy preservation practices are unnecessary. Other methods result in food being able to be stored for months and even years at a time, making them excellent options for long term storage. That is precisely what you want to achieve while in quarantine as it allows you to store food for as long as possible, effectively moving your reliance away from finicky supply chains and into your very own pantry.

How food is preserved depends on what type of food is being preserved and how long you need to preserve it. It also depends on what you have available to you. Not all equipment that is meant for preserving comes cheap or easy, which means that some options may not be available. With that being said, every option *can* be done in your own home; it's just a matter of how much space you have and how much you are willing to invest in the preservation process to get there.

What Types of Preservation Methods Are There?

There are many different types of preservation methods out there! There are also many different types of storage methods. Note that preservation and storage methods are not always the same. For example, when you are canning food, your preservation method and storage method are pretty much one and the same. When you dehydrate food, dehydrating will prepare it for preservation; however, you will need to use an alternate method to store it properly. Vacuum sealed packages are most common for dehydrated foods, especially meats. Sterilized and tightly sealed jars may be used for other types of dehydrated foods, such as dehydrated herbs or vegetable bits.

The most common preservation methods that people use include, water bath canning, pressure canning, dehydrating, freezing, brining and salting, sugaring, smoking, pickling and fermenting, and using ash, oil, or honey to preserve things. Each of these preservation methods will ensure that the food you are preserving is clean, sterile, and shelf-stable so that you can safely store it in your pantry or kitchen for an extended time.

Once you have preserved something, you will need to know how to prepare it after properly. Some things can be served straight out of storage, while others will need to be prepared first. For example, jam or jelly can be eaten right out of the jar, however canned potatoes will still need to be cooked, or canned soup will need to be reheated. Different types of preservation will require different levels of preparation, too. This will especially depend on what has been preserved and at what stage. For example, frozen meat will still need to be defrosted and cooked before consuming because the average freezer does not get cold enough to safely kill off any bacteria and parasites that may be living in raw meat.

How Do You Choose Which Preservation Method to Use?

Choosing which preservation method to use will depend on a few things: what food you want to preserve, what you want to do with the preserved food, when you want to use it, and what tools you have on hand.

Not all preservation methods are made equally, which means not all preservation methods are suitable for all types of foods. A great example would explain the two different types of canning: water bath canning and pressure canning. While water bath canning is excellent for more acidic foods such as jams, jellies, tomato sauces, and certain fruit juices, it is not ideal for less acidic foods such as vegetables, soups, or meats. This is because a water bath canner simply does not get hot enough to safely sterilize the jar and the contents of the jar during the canning process. Pressure canning is used for less acidic foods such as vegetables, soups, and meats because the pressure canner is able to get extremely hot. That heat penetrates the jars and ensures that everything inside of them is sterilized properly.

Aside from which type of food you are planning on preserving, you need to consider what it is that you want to do with that preserve. Most foods will be able to be preserved in a few different ways, so you need to decide what your goal is with the final product. Meat, for example, should be preserved based on how you want to prepare the meat afterward. If you are planning on cooking and incorporating it into a recipe, smoking your meat may not be ideal, unless your recipe specifically requires you to have smoked meat. Consider what the preservation process will do to the ingredient you are preserving and preserve your food based on what you plan on making with that food item in the long run.

Knowing when you want to use preserved food matters, too. Most preservation methods will vary in how long they can safely be consumed for. For example, berries can be frozen for about one year before they begin to lose quality and stop tasting as good. Alternatively, canned fruit can last up to 2 years. Consider storing one type of food in a couple of different ways so it can be used in different types of recipes and for differing lengths of time. This way, you have plenty to last you for a long time, and you can enjoy a varied diet.

Lastly, you need to consider what you have on hand. Especially in the middle of a pandemic where supply chains are damaged, it may be challenging to get exactly what you need in order to preserve food in specific ways. Before you commit to preserving one type of food in any particular way, consider which preservation tools you can get your hands on and then go from there. If you can easily access and afford all of the tools for any chosen method of preservation, then you are good to go. Otherwise, you will need to choose an alternative method that is more accessible and affordable for you.

Are All Preservation Methods Easy Enough for Beginners to Try?

Not all preservation methods are easy, though a beginner can certainly try any method he or she wishes to try. With that being said, it is important that you do not try a recipe that you are not confident in, as you could find yourself making mistakes based on a lack of understanding. These mistakes can lead to the development of harmful bacteria or parasites, which, in turn, can spoil your food and cause serious illness in anyone who consumes the foods.

Always make sure you are confident in the method you are using and that you clearly understand it so that you can safely prepare food for your family. If you are ever uncertain, see if you can recruit the help of someone who knows what they are doing. Watch some YouTube videos which can provide you with more in-depth guidance on each method, or choose an alternative method for preserving your food. As you go, you will find it easier for you to preserve your food, which will help with building your confidence and will support you in adding even more to your home prepared food storage.

Things That Will Destroy Your Preserves

Before you jump into preserving food for yourself and your family, you need to know which things are going to destroy anything you have preserved. Knowing this in advance ensures that you can avoid or minimize these exposures so that you can keep your preserved food for as long as is possible.

The top six things that will destroy your preserves include light, oxygen, moisture, temperature, pests, and time.

Light can create two unwanted situations which are both responsible for damaging your preserved foods. UV radiation has been known to damage certain components of certain foods, causing those foods to break down prematurely, which results in them losing their flavor, texture, and nutritional value. While these foods might fill your stomach, they will not be nearly as enjoyable to eat, and they will not offer as many nutrients as they could have otherwise. Electric lights can still reduce shelf life by increasing the heat that surrounds the food, which can cause it to deteriorate, too. The other element of light that can damage your food is the fact that sunlight can increase oxidization. As the food inside heats up, it begins to produce more gases on its own, which results in more oxygen reaching the food, effectively minimizing its shelf life and possibly spoiling the food altogether.

Oxygen reduces the quality of preserves, possibly to the point of completely spoiling them, in a process known as oxidization. Oxidative spoilage destroys foods that have fat content in them because the lipids oxidize, which results in short-chain carbon compounds being formed. These compounds have a strong odor and flavor and can encourage the development of unwanted bacteria in the food itself.

Moisture reduces the quality of preserves as it can develop in the creation of mold and other spoilages that destroy the food. If moisture gets into food, all sorts of bacteria can develop on or in that food, rendering it dangerous to eat. If you were to eat food that was damaged by moisture in storage, you could fall extremely ill from the bacteria that grew as a result of the moisture.

Temperature reduces the quality of foods when the temperature is too hot or too cold. Too hot can result in the formation of "sweating" inside of the product itself, which can create moisture and effectively cause the development of bacteria. Too cold can cause the item to get freezer burn and can destroy the flavor and quality of the food itself. The ideal temperature for storing a preserved food item depends on what method you have used. However, all foods will not tolerate being kept too hot or too cold.

Pests can completely destroy the quality of preserved food. When a pest has gotten into your food, it is best to throw the food out entirely as pests can bring with them many different risks when they have contaminated your food. Pests can bring about bacteria and viruses by way of feces, urine, and saliva, and they can infest your food with parasites. You should never consume any food item that has been affected by pests as you will be running high risks. Once bacteria or parasites have been introduced to your food, they can multiply rapidly. You do not want to find yourself consuming food that has become contaminated in such a way as the results can be massive, possibly even fatal.

Lastly, time will reduce the quality of your food preserves. Most preserved food will last for many years, possibly even a few decades. However, as time goes on, you will find that those foods start to deteriorate in quality. Eventually, enough time will pass, and the food will no longer be consumable simply because of how much time has gone by. While proper preservation can prevent food from breaking down, eventually, everything breaks down.

Chapter 2

Top Foods to Stockpile

Knowing what food to preserve is essential. There are a few things that will factor into what types of food you should be preserving for yourself and your family. It is important to have a plan for what you will preserve, how much of it you will preserve, and how you will preserve it clearly defined in advance. Going into the process of preserving food without a clear plan can result in you not having enough, having too much of one thing but not enough of another, or having more than you actually need. Not having enough or not having enough of a variety can lead to a poor quality diet, which can ultimately deteriorate your health and put you at greater risk of falling ill. Having more than you actually need can waste your money, which is never a good idea. While it would be better to end up with more than you need, it would be unfortunate to waste a precious resource such as money when this resource is already stretched very thin for most people.

The Master List of Foods to Preserve, Including Superfoods

This list contains all of the foods you should look at preserving for your family, including superfoods that are known for having an abundance of nutrients in them. In an ideal world, you would be able to preserve every single item off this list. However, that may not be reasonable for you as some items may not be easily accessible or affordable, or they may be things that your family simply won't like. We will discuss how you can decide which foods you will preserve for your family next. In the meantime, look through the following lists below to get an idea of what foods you can and should preserve from home.

Meat and Dairy

- White Tuna
- Salmon
- Oyster
- Mussels
- Lobster
- Crab
- Clams
- Beef, Venison, Elk
- Pork, Bear
- Goats, Sheep
- Rabbit
- Pilchards
- Bone Broth
- Eggs
- Milk
- Butter

Vegetables

- Herbs (cilantro, dill, marjoram, pepper, etc.)

- Tomatoes
- Sweet Potatoes
- Carrots
- Green Beans

Fruits

- Mango
- Apples
- Pomegranate
- Berries (strawberries, raspberries, blackberries, etc.)
- Bananas

Miscellaneous

- Hazelnuts
- Dark Chocolate
- Extra Virgin Olive Oil
- Kefir
- Miso
- Raw Honey
- Whole Grain Bread
- Peanut Butter

Pantry Supplies

- Dried Pasta (macaroni, spaghetti, fettuccini, etc.)

- Rice (brown rice has more nutrients, white rice lasts longer)
- Powdered Dairy Products
- Oats
- Sugar (white, brown, syrup, molasses, honey)
- Salts
- Leavening Agents
- Potato Flakes
- Water

Superfoods

- Broccoli
- Garlic
- Kale
- Licorice
- Citrus Fruits
- Bell Peppers
- Ginger
- Spinach
- Yogurt
- Almonds
- Turmeric
- Green Tea
- Papaya

- Poultry
- Kiwi
- Watermelon
- Blueberries
- Acai Berries
- Elderberries
- Sunflower Seeds
- Button Mushrooms
- Medicinal Mushrooms
- Astralagus
- Pelargonium Sidoides
- Spirulina
- Moringa
- Oysters
- Shellfish

Saving Costs Through Preserving Foods

Preserving foods may seem expensive if you look at the up-front cost. Still, the reality is that if you break that cost down over time, you will discover that the overall price is much lower than your average grocery bill. Further, there are many things you can do to help you minimize the costs associated with preserving food.

First and foremost, you need to realize that the "available all the time" mentality is something that was instilled in us by the very supply chains that are presently having a hard time managing our current global affairs. Commercial greenhouses, unnatural growing situations, and access to a worldwide market all resulted in suppliers being able to offer you any food you want, whenever you want, and often at a fairly reasonable price. This is not how it "should" be, though, and when it comes to preserving your own food, following this system can be costly. It can also destroy the flavor and quality of your preserved foods by resulting in you preserving food that is low quality, to begin with, or that may be filled with harsh chemicals that were used to make it grow.

Learning how to buy with the seasons means that you can buy your food as it is in stock from local suppliers. If you can preserve that food as soon as possible, there is a very short window from the item being harvested to the item being preserved. Doing this ensures that your food is the highest quality, is as fresh as possible upon being preserved, and that it lasts much longer. Your food is going to taste better and cost far less, as well.

In addition to buying your foods in season, buy them from local growers and farmers. Local growers and farmers will often offer you the best prices on supplies, especially if you are buying them in bulk, because you are sending so much business their way. Plus, you will be stimulating your local economy. You will know exactly where your food has come from and how it was grown, which are two significant benefits when it comes to surviving a global crisis such as the one we are in right now.

Other ways to save money on sourcing your food for preservation comes from knowing where you can generally go to buy produce and groceries for cheaper. These places are well-known for offering great deals and discounts, all of which can help you save more money while preserving food for yourself and your family.

The following list of resources is sourced from "A Year Without The Grocery Store: A Step by Step Guide to Acquiring, Organizing, and Cooking Food Storage" by Karen Morris. This shares all of the best places to buy your supplies, with the cheapest possible price tag on them:

- Online co-ops
- Bulk food stores and the bulk section of regular stores
- Online food storage companies
- Latter-Day Saints canneries
- Restaurant supply stores

You can also cheapen your items by couponing and by buying in bulk as much as possible, as this often creates a situation where you are paying less per item.

Chapter 3

Effective Stockpiling

How to maximize nutrients, focus on staples with long shelf life, meal prep and preserving leftovers, examples of best foods to stockpile, a mention of what to do if you suffer food allergies or specific health issues relating to food (i.e., Chron's disease, diabetes, etc.)

Determining What Types of Food Your Family Should Preserve

Before you choose what types of food you should preserve, you need to consider what the needs of your family are. Consider your nutritional needs, your nutritional preferences, and any food allergies or intolerances anyone in the family might have. Then, consider the amount of money that you can reasonably budget toward building up your food preserves to ensure that you are getting the best nutritional value that is right for your family, and that fits within your budget.

Start by going through the list above and highlighting every food your family can and will eat. This way, you narrow the list down based on what is reasonable for you to preserve. Next, you are going to consider nutritional value. You are going to want to have at least 3-4 different types of protein stored, 8-10 different types of vegetables stored, 5-7 different types of fruits stored, and a pantry stocked with everything you will need to turn all of your foods into proper meals and get you through the foreseeable future. Highlight several items from each category and favor items that are known for having higher levels of nutritional value. This will give you well-rounded food storage available for you at all times.

Determining How Much Food You Need to Preserve for Your Family

Lastly, you need to consider how much food you need to preserve for your family. This part can be challenging as you are going to need to do some calculations to identify how much food your family needs to sustain themselves for a year.

The absolute minimum amount of recommended food to be preserved per person per year is:

- 365 pounds of vegetables
- 321 pounds of fruit
- 300 pounds of grain
- 182 pounds of meat
- 75 pounds of dairy
- 75 pounds of legumes
- 65 pounds of sweetener
- 5 pounds of salt
- 4 pounds of shortening
- 2 gallons of oil
- 2.5 pounds of leavening agents (yeast, baking soda, baking powder, etc.)

This may sound like a lot, but keep in mind that you do not *have* to store enough food for up to a year. You can store enough food for 3 to 6 months to start and go from there. You can always add more to your inventory every few months to keep you going this way for the long haul. Not everything needs to be added at once, either, nor should it be. Starting with shorter periods may be ideal because it will allow you to get used to storing and using your food preservation stores.

Chapter 4

Planning and Acquiring Your Foods

Properly planning and acquiring your foods ensures that from the very start of your food preservation process, you are organized and ready to preserve everything properly. A proper plan in place ensures that you know what you need to do every step of the way, so that you can acquire all of the right ingredients and supplies for preserving, and that you preserve your food quickly. Once you have obtained your food, you want to preserve it as soon as possible so that it is fresh during the preservation process. This ensures that you do not end up with wasted food that went bad before you were able to get to it. As well, having a clearly organized plan ensures that you do not skip or miss any steps due to being overwhelmed, which means the preservation process will be done properly, and you will not be at risk of getting sick.

Creating a List of All of Your Foods and Supplies

You want to start your planning process by first writing a master list of all of the foods and supplies you are going to need to acquire your foods and preserve them properly. Starting with the food is easiest. Then, research the preservation methods you are going to use and compile a list of all of the tools you are going to need to preserve your food properly. If this is your first time preserving, your food list may be rather large and expensive because you will need to buy things that are not necessarily consumables, such as a pot or a pressure cooker for canning, a smoker, or a vacuum sealer.

Be sure to put every single thing on the list so that you know what it is that you are going to need in order to preserve your food properly. This is going to help you plan your budget and schedule around how you are going to preserve all of your food storage properly.

Once you have added every single thing to your list, estimate roughly how much each item is going to cost you. Knowing how much things will cost will help drastically when it comes to budgeting and planning your schedule. It will allow you to do things based on when you can reasonably afford to do them.

Planning What Will Be Preserved and When

Next, you need to plan what will be preserved, and when. Again, you want to preserve things when they are in season as they will be cheaper *and* fresher, which means your preserves will be higher quality and they should last longer.

The best way to plan out your preserved foods is by what will be preserved and when you need to schedule yourself on a month to month basis for the next 12 months. Write the schedule out based on what types of food will be in season and available for purchase during those months. This will allow you to identify how you can follow the natural harvesting seasons.

Here is a sample of what a preserving season looks like:

- January – April: Winter squash, garlic, peas, beans, carrots grown indoors
- May: Rhubarb, eggs, herbs, winter squash
- June: Garlic, strawberries, snap peas, shelling peas, herbs, onions, fish
- July: Berries, beets, root vegetables, summer squash, beans, fish

- August: Tomatoes, peppers, watermelon, corn, fish
- September: Rhubarb, eggs, herbs, garlic, strawberries, snap peas, shelling peas, herbs, onions, berries, beets, root vegetables, summer squash, beans, tomatoes, pepper, watermelon, corn (most crops are in second harvest by this time)
- October: Apples, pears, quinces, meat
- November: Cabbage, carrots, bok choy, spinach, potatoes, onions, meat, nuts
- December: Organize the root cellar and plan for the next year

Budgeting for Your Preservation Plan

Now that you know what needs to be preserved and when you need to plan your budget. This is going to give you the "final say" in how your preservation season will go since you will need to be sure that you can reasonably afford to preserve everything you have chosen.

Because every month is going to have such different harvests and preservation needs, you are going to want to know how much each month will cost you. Consider both the foods you will need to buy *and* the resources you will need to make that happen. This way, you know an exact calculation of what that months' worth of preservations would cost you.

Once you know the cost of preserving, you need to factor it into your annual budget. If you have not already made one, now is the time to start. See if the cost for preserving as much as you need and want is feasible and, if it isn't, see where you can trim costs. You can rely on the resources mentioned in *Chapter 2: Top Foods To Stockpile* as a way to cut costs. You can also go to local farms and farmer's markets, where direct suppliers will often offer great discounts on items that are ready to be sold. In some cases, you may be able to find some items that are nearing the end of their shelf-life in fresh form for cheaper because they are nearly ready to be tossed. These types of foods are often good for canning because canning will preserve them for much longer.

Another way you can adjust your canning season to fit your budget is to swap out more expensive items for cheaper ones. For example, lobster and crab can be quite expensive, so you might swap them out for salmon and tuna instead. Certain berries like elderberries also tend to be more expensive, so you might consider swapping them out for blueberries. Making these types of swaps in your preservation plan creates a cheaper preservation process, effectively helping you save as much money as possible.

Don't just look at saving money on the food itself, either. Often you can find resources on sale or even available at lower prices second hand. Just make sure that anything you buy, whether it is new or used, functions properly, and can create the results you desire.

Acquiring All of the Resources You Will Need First

Before you begin your preservation process, you are going to want to acquire your resources first. Look into what types of preservation methods you will be doing in a given month and stock up on everything you will need for those methods. If you are canning, make sure you have a proper stockpot, water bath canner, jar holder, jars, pectin, and anything else you need for canning. If you are going to be smoking your food, make sure your smoker has fuel, you have the proper wood chips, and you have a vacuum sealer with fresh bags to use to help you store your smoked meats after.

Before you go to purchase your food supplies, make sure you double-check all of your resources. You want to be sure that you have everything you need *and* that everything you have is in proper working order. This way, you do not find yourself without something partway through making a recipe and are forced to either stretch yourself thin to get it or screw up your entire preservation recipe because you do not have it. Remember, with most preservation methods, you will not have the option to substitute various things because doing so can mess up the preservation process and introduce harmful bacteria into your foods. You must have exactly what you need and in perfect working order before you bring your food home so that you can safely preserve your harvest every single time.

The other resource you need is one that will likely not cost anything unless you choose to invest in a cookbook. That is, you are going to need recipes. Recipes can be found online or in cookbooks that are designed for your preferred method of preservation. Be sure to rely on tried and tested recipes and not just any recipe, as the recipes that are truly intended for preservation will be written with safety considerations in mind. Do not purchase any food supplies until you have compiled all of the recipes you will be making, as this will ensure that you are able to create a complete grocery list beforehand and that you can go out and get these items all at once. That way, you are not left running back and forth to the store, trying to pick up everything you forgot about the last time you went.

Bringing Home Your Food Supplies

Once you are sure you have all of your preservation supplies and resources available, you are going to need to bring home your food supplies. Bringing home your food supplies should be done in a three-part process.

First, you are going to want to locate where you are going to store your food supplies for the short term as you process everything. It may take a few days for you to process everything completely, so you need to make sure you have room on your shelf, in your fridge, or in your freezer to hold onto the excess supplies until you are able to get to it. If you do not know where you will be storing your food yet, do not bring it home right away as it could lead to you destroying the quality of the food before you can safely get it into your preservation system.

Second, you are going to want to source where you are getting your food from. At this point, especially if you're going to be as frugal as possible, you are going to need to shop around to make sure you get the best deal. Look around to see what is available and find the best deals possible; if you can, factor in the cost of sale pricing, coupons, and other deals that may help you save more money. If you are going to be buying more than one thing in a month, make sure you consider the convenience factor of picking everything up. You don't want to drive around to several different stores and locations with your food sitting in your vehicle, especially in warm months. This will waste time, gas, and possibly cause your food quality to deteriorate while it sits in the car. Try to get everything from one, two, or maybe three different locations at most.

Third, you are going to have to get your food home. For the most part, you should only need to stack everything in your vehicle. If, however, you do not have a vehicle or if you have more than you can reasonably put in your vehicle, you are going to have to find an alternative way to bring your produce home. When pandemic orders are not in effect, you may be able to rely on an Uber or taxi to help you. When pandemic orders are in effect, it would be safer to rely on a family member or a friend or to have the store deliver your food to your house if possible. However, that will cost more as you will have to pay a delivery fee for your food.

Creating Your Recipe Plan

If you are going to be processing all of your food over a day or two, you can create your recipe plan once your food is at home. If you are going to be processing food every week for a month, you are going to want to do this ahead of time so that you only buy the food that you will be able to process in the next couple of days after bringing it home.

Your recipe plan essentially states which recipes you are going to use and when you are going to use them in the day. Having a detailed recipe plan ensures that you stay organized on processing day, so you do not lose track of what needs to be done or what has already been completed. I suggest organizing your recipes based on the major ingredients in them, or the types of recipes you are doing so that you can keep it clean and organized. For example, do all of your recipes with tomatoes in a single day or over two days; that way, you are able to process all of your tomatoes and have them done finished. If you are going to be using two major ingredients in a single day, work with one first, and then the next.

Keeping your recipes organized means you do not have to keep putting things away and taking them out again to do your next recipe, and then your next one, and so forth. This way, you save time and can move from one recipe to the next quickly, and then you can clean up all at once. Or, on a particularly large day, you may want to plan for cleaning periods in between your recipes so that you can bring the organization back to your kitchen. Remember, disorganized kitchens can be a hazard for cuts, burns, and other injuries, so you want to keep your kitchen as clean and calm as possible.

Executing Your Food Preservation Plan

Lastly, you need to execute your food preservation plan! Once you have made one, it is important that you do not stray away from it and that you stay on track with the plan you made for yourself. Only make what you put on the plan, and in the exact order you planned. This way, you do not end up running out of ingredients or finding yourself confused about where you are in the plan and missing important recipes.

The only time you should alter your plan is if you run into trouble that seriously stops you from being able to proceed with a certain recipe or ingredient. In this case, you should stop immediately (or as soon as you can if you are mid-process) and decide on a new plan that factors in safety and efficiency. Otherwise, everything should be done per plan. If you do run out of time in one day, simply carry on the next day.

Chapter 5

Storing Different Types of Food

As I mentioned previously, every single type of food has its own methods for being stored. Storing food properly for the type of food it is, is essential because this is how you are able to prevent that food from spoiling. It is imperative that you never attempt to preserve or store food in a method that is not approved for that variety of food because you are running the risk of allowing harmful bacteria, such as *Clostridium Botulinum,* or the bacteria that causes Botulism.

Below, you will discover the best methods for storing each different type of food safely. Again, you are going to want to pick which kind of preservation you use based on what you plan on using that ingredient for later. For example, if you're going to be able to cook a steak, later on, turning all of your beef into beef jerky is not the right choice. Consider how you intend to use your harvest before preserving it so that you can use it for its intended purposes later on.

Meat

Meat can be preserved in a variety of ways, and how you preserve it will affect the taste, texture, and how it can be used later on.

The following ways are approved for storing different types of meats:

- Smoking:
    - This method is most common in red meats, though it can be done in poultry, too. It will create a drier, chewier texture in the meat itself. You

will also notice the flavor of the wood that was used to smoke it. Smoked meat should be consumed sparingly as the smoke itself contains carcinogens.

- Salting:
    - Salting, or curing, is an old school preservation method that can still be used. Salt requires more time and effort, but it is cheap. It is often used for meats like bacon or pastrami. You can also salt then smoke meats as a way to enhance flavor.
- Brining:
    - Brining is another traditional preservation method. Brine is made of water, sugar, and salt, and it is poured over meat as the meat is held down at the bottom of the brine using a weighted crock. This method can also take up a lot of space.
- Canning:
    - You cannot water bath can meat, but you can pressure can it. Meat is not acidic enough for water bath canners. Any type of meat can be pressure canned, and this method is most simple as it requires no further work to preserve it. When you are ready to consume the meat, you open the jar and reheat the meat itself. Canned meat is cheap to make, and the preservation process will turn even tougher meat cuts into tender, delicious meat.
- Dehydrating:
    - This method is easy, cheap, and healthy. Electric dehydrators or solar dehydrators can be used. You should buy the largest dehydrator you can

afford, so you are not running several batches through a smaller dehydrator. Dehydrated meat can be eaten as is. It will need to be stored in vacuumed sealed packages until you are ready to eat.

- Stored In Lard:
    - If you are storing an animal with a lot of fat, like a pig, you can store it in a crock with lard. This method works the same as brining, except you are using lard instead of brine. The lard completely surrounds the meat and prevents air from getting to it. It is cheap and effective and requires no special equipment.
- Freeze Drying:
    - Freeze driers are expensive, so this is not a practical method for most people. However, freeze-dried meat can be stored for a long period of time, it is lightweight, and the food retains nearly all of its nutrition. These foods are also great if you need to evacuate in case of an emergency because they are so light. Freeze-dried meat can be rehydrated by adding it to a liquid and cooking it, though the texture will be very different from other meats.
- Freezer:
    - The freezer is a great way to store meat. Make sure you store it in heavy-duty freezer bags or containers so that it does not become freezer burnt. If meat does become freezer burnt, however, it can be used to make a stew or other similar recipes, so long as the freezer burn is not too bad. Frozen meat stays closest to its raw form, which means it can easily be dehydrated and incorporated into virtually any recipe.

Seafood

Seafood needs to be preserved carefully as it is more finicky and more prone to bacteria and parasites that are unsafe for humans. It is important to properly clean and preserve your seafood to avoid contamination or getting anyone sick in the process.

The following methods are approved for safely preserving seafood:

- Freezing:
  - Freezing is generally the most popular method for storing seafood. For lobster, freezing is the only safe way to store seafood. Seafood tends to be especially prone to the development of bacteria, and the meat is so delicate that preserving it in other ways can damage the meat. There are, of course, exceptions to this, however.
- Brining:
  - Seafood can be brined in salt before being stored. Varieties like herring, salmon, rockfish, and mackerel can be brined and kept for up to 9 months using this method.
- Drying:
  - Drying is a popular way of storing fish, and it can be stored for up to 2 months this way. Most varieties of fish can be dried. Once a fish has been dried, it can be crushed as a condiment or rehydrated by adding it to a soup recipe.
- Smoking:

    - Smoking a fish is a great way to preserve it while creating a delicious texture and flavor. Most fish can be smoked, though salmon and oysters are particularly common varieties for smoked seafood options.
- Canning:
    - Fish should only ever be canned in a pressure canner as water baths will not safely can them for you. Fish are not acidic enough for water bath canners. Virtually every variety of fish can be canned except for lobster. Canning a fish will make it incredibly juicy and will make it delicious to enjoy later on. Plus, it retains its ability to be used in a variety of recipes this way.

Vegetables

Vegetables are fairly easy to preserve, though different varieties will require different types of preservation methods. Leafy greens, for example, are hard to preserve because they are so full of water content and delicate that they can quickly deteriorate in most storage methods.

The following methods are approved as safe ways to preserve your vegetables:

- Drying:
    - Dehydrating vegetables removes the water content from them so that they can be safely stored. You need to make sure dehydrated vegetables are completely dehydrated so that they do not develop bacteria. Food needs to be at least 95% dehydrated before it is considered shelf-stable, and it will still need to be stored in a proper container.

- Canning:
  - Like meat and seafood, vegetables are not acidic enough to endure water bathing. You will need to use a pressure canner to can your vegetables properly. Most vegetables can be stored this way safely for quite a long period of time. Note that vegetables stored this way will be quite moist, so you will not be able to retain the fresh, crunchy texture this way.
- Pickling:
  - Pickling vegetables requires you to have an acidic base such as vinegar to store your vegetables in. Not all vegetables will taste right when they are pickled, though you can pickle a wide variety of things. Cucumbers, asparagus, bell peppers, beets, carrots, cauliflower, fennel, ginger, green beans, mushrooms, onions, parsnips, hot peppers, radishes, rhubarb, squash, and turnips can all be pickled, among other things.
- Fermenting:
  - Fermenting is a process where you convert carbohydrates to alcohol or organic acids. You do this using salt, whey, or another starter culture to a vegetable as a brine and letting the food sit in the brine so that it ferments. Fermented foods are incredibly healthy, as long as they are done right, and they are quite easy to do. Most foods should be able to be fermented relatively easily.
- Freezing:
  - Freezing is a good way to retain the freshness of your vegetables. When you freeze vegetables, you may need to blanch them or flash freeze them in order to freeze them effectively. In some cases, you may need to do both.

- Oil Packing:
    - Oil packing vegetables means that you are essentially storing vegetables in oil. Oil creates anaerobic conditions, meaning that virtually no air can get to the vegetables inside, which results in them being stored safely. This process can be used on tomatoes, eggplants, herbs, onions, and olives.
- Salting:
    - Like meat, vegetables can be salted to cure them, too. Salting vegetables is done in a different means as salting meat because the content of vegetables is so much different. However, it does cure and preserves them all the same.

Fruit

Fruit is similar to vegetables in that it is fresh and grown out of the ground. However, fruit is more acidic than vegetables are, which means fruits can be preserved in a couple of different ways, too.

The following methods are approved as safe methods for preserving fruit:

- Drying:
    - Like vegetables, fruit can be dehydrated. Dehydrated fruit can be consumed as is, it can be added to recipes, or it can be rehydrated and consumed that way. However, rehydrated fruit will not taste the same as it did before it was dehydrated.
- Canning:

    - Fruit is one of the only things that can be done safely in a water bath canner, aside from some recipes that are meant for pickling. With fruit, canning can be used to preserve the whole fruit, or it can be used to allow you to break fruit down into other things such as jams, jellies, juices, salsas, and sauces, which can then be canned and stored.
- Pickling:
    - Believe it or not, fruits can be pickled! Blueberries, cherries, grapes, mangoes, peaches, strawberries, tomatoes, and watermelon can all be pickled and stored.
- Fermenting:
    - Fruits are less likely to be used in ferments as vegetables are unless you are making wine, of course. Fruit can be fermented, however. It typically takes much less time to ferment fruit because it is already acidic. It can take up to 48 hours to ferment fruit, while it can take several days or even a couple of weeks to ferment vegetables.
- Freezing:
    - Freezing is a great way to store your fruits. Most fruits will freeze well. Fruits will not need to be blanched before being frozen; however, it can benefit to cut them up and flash freeze them first. Certain fruits like citrus fruits, tomatoes, bananas, and watermelon do not freeze well because they become mushy and gross once they are defrosted. If, however, you only plan on using them in smoothies or in baking recipes, it may not be such a big deal to freeze them. However, there are better means of storing them.

Dairy

Dairy products can be preserved in a variety of ways. With that being said, not all dairy products are the same, so you are going to need to preserve them properly to be able to use them. As well, you are seriously going to need to consider what you want the dairy for, as most preservation methods will alter the flavor and usages drastically.

The following are approved and safe methods for preserving and storing dairy products:

- Freezing:
    - Freezing is the most dependable way to extend the shelf life of dairy products, including eggs. With that being said, once you freeze a dairy product, it will separate, and it will never taste quite the same, even if you stir it up afterward. It will still be good for baking, or for survival means, however.
- Dehydrating:
    - Eggs can be dehydrated in a food dehydrator and, when done this way, can store for months. Cheese can also be dehydrated. Eggs will need to be recooked before they can be used, while cheeses can be crushed into a powder and dusted on top of foods for flavoring means.
- Wood Ash:
    - Wood ash will increase the shelf life of both eggs and cheese. Eggs stored in hardwood ash can last for up to a year, while firm cheeses can be stored for about three months this way. Note, however, that cheeses will start to taste like the ash after about three months.

- Oil Storage:
    - Softer cheeses like mozzarella and feta cheese can be submerged in olive oil and stored for quite a while, although they will still need to be refrigerated. If you oil eggs first, they can be stored at room temperature for up to 5 weeks.
- Wax Storage:
    - Hard cheeses like cheddar can be coated in wax and stored outside of the fridge in a cool, dry place.

Bulk Foods

- Air Tight Containers:
    - Most bulk foods just need to be stored in sterile, airtight containers. For smaller amounts of an ingredient, you can use sterile mason jars. For larger ones, you can invest in larger containers that are usually used in grocery stores or restaurants. Containers are usually made from HDPE or high-density polyethylene and are easy to sterilize and keep airtight.

Dried Foods

- Air Tight Containers:
    - Like bulk foods, airtight containers are often the best way to store dried foods. Sterile mason jars or larger bulk containers can be used for any dried food.
- Vacuum Sealer:

- Vacuum sealers are great for dried foods, especially those that may still have a small amount of moisture in them, such as raisins or other dried fruits or meats. Vacuum sealers are the best way to ensure that absolutely no air gets into your food so that you can keep it stored and safe for a long period of time.

Chapter 6

Canning

Canning is the most popular method of storing, aside from freezing, for the average household. Canning is easy to get into because the materials are quite affordable, they are abundant, and these types of foods can last for an extended period of time. Another benefit of canning is that so many foods can be preserved this way, which means you do not have to invest in several expensive materials to get started. If you are new to preserving food, getting started with canning is a great idea as it allows you to jump right into preserving many different types of foods. Plus, the mason jars you buy for canning can also be used for preserving other foods such as dried ingredients, too.

If you are going to get into canning, you should start with a pressure canner. You can always purchase a water bath canner at a later date, but pressure canners will do far more varieties of food for you. These days, there are electric pressure canners available which tend to be easier to manage if you are new to canning because they automatically control their temperature, you do not have to do it for them. However, these pressure canners are quite a bit smaller, so it will take longer for you to get everything done.

The methods for canning are similar, but there are some key differences between the two styles. Be sure to research the specific method of canning you will use beforehand and educate yourself on how that method works before getting started so that you do not injure yourself or damage your food in the process.

Water Bath Canning

Water bath canning is a popular canning method among hobbyists who like to make jams, jellies, salsas, and other similar recipes. A water bath canner is a large pot that holds at least 7-quart jars while allowing them to be safely submerged in water with 1-2 inches of water above the heads of the jars. The canners have a rack that is submerged in water that holds the jars, so they are not directly against the bottom of the pot by the heat source. The jars then boil for a certain amount of time, based on what recipe is inside of the jar and what altitude you are at.

What Foods Does Water Bath Canning Work for?

Water bath canning works for most fruits, as well as fruit juices, salsas, tomatoes, pickles and relishes, chutneys, sauces, pie fillings, vinegar, and certain condiments.

What Materials Do You Need for Water Bath Canning?

In order to water bath can you will need:

- Mason jars with new lids (or reusable lids. You cannot reuse non-specified reusable lids as they will not reseal properly.)
- Water bath canning pot
- Jar lifter
- Large stockpot
- Stir stick
- Timer
- Canning funnel
- Pectin

How Does The Water Bath Canning Process Work?

Water bath canning works by first preparing a recipe to place inside of your jars. For jams, jellies, salsas, chutneys, sauces, and so forth, you will need to prepare a specific recipe for your canning method. For other things, such as storing sliced fruit, you will need to make a specific syrup to store the fruit in.

Once you have made your recipe, you will need to place that recipe inside of your sterilized jars, add the lids, and boil them in the water bath canner for a specific period of time that's based on the recipe inside and your altitude. This will kill any *Clostridium Botulinum* bacteria inside of the jars so that everything within them is safe. You will remove the jars afterward using a jar lifter and then place them on your counter. Then, you will leave them on the counter overnight to cool. You will check the lids in the morning to make sure they are all properly sealed. Do not tighten the rings at this point; just leave them as they are. If any lids pop off, you will consume those right away. Others can be stored on the shelf for 1-2 years.

Note that with canning recipes, jars can be smaller than defined in the recipe but not larger. The amount of time that jar spends in the water bath will remain the same unless otherwise noted.

Simple Strawberry Jam

8 x 250mL jars.

What You Need:

- 5 cups Crushed Strawberries

- 7 cups Granulated Sugar
- 4 TBSP Lemon Juice
- ½ TSP Butter
- 1 Pack Pectin

How to Make It:

1. Place eight sealed jars in your water bath canner. Fill with water to cover the canner with 1-2 inches of water. Remove the lids, set the screw bands aside, and drop the sealing discs into the hot water, but not on top of the jars. Simmer at 180F as you prepare your jam. (This will sterilize the jars.)
2. Prepare strawberries and crush them one layer at a time. Measure out 5 cups of crushed berries. Measure and set aside granulated sugar.
3. In a pot, combine strawberries, lemon juice, and butter. Dissolve pectin into the mixture.
4. Bring the mixture to a full rolling boil, then add all of the sugar. Stir constantly while returning the mixture to a full rolling boil that does not disappear as you stir it. Let it boil for 1 minute while stirring. Remove the entire mixture from the heat. If foam forms, skim it off.
5. Remove jars from the water and carefully use a tea towel to dry the jar, being cautious not to burn yourself. Ladle the jam into the jar, leaving ¼" headspace. Use a nonmetallic utensil to remove air bubbles. Adjust headspace if needed by adding more jam. Wipe the rim clean. Remove a sealing disc from the water, dry

it, and carefully center it over the jar. Screw the screw band on until it is fingertip tight. Return the jar to the canner. Do this with all of the remaining jam.

6. Ensure the jars are covered with at least 1 inch of water. Cover the canner and bring it to a full rolling boil. Boil the filled jar for 10 minutes for altitudes of up to 1,000 feet.
7. When done, remove the lid, turn off the heat, and wait 5 minutes. Without tilting them, remove the jars from the canner and place them on a layer of two tea towels on the counter, so they don't scald your countertop—cool upright for 24 hours. Do not retighten the screw bands.
8. Check seals after 24 hours. Sealed discs will curve downwards and will not move at all if you press on them. Remove screw bands, wipe and dry them and the jars, and replace the screw bands on the jars loosely. Label and store your jars.

Tomato Salsa

Makes 10 x 250mL jars or 5 x 500mL jars.

What You Need:

- 7 cups Tomatoes, chopped
- 2 cups Onions, chopped
- 1 cup Green Bell Pepper, chopped
- 3 Cloves Garlic, minced
- 8 Jalapenos, chopped
- 156mL Tomato Paste
- 175mL White Vinegar (3/4 cup)

- ½ cup Cilantro, chopped and lightly packed
- ½ TSP Cumin, ground

How to Make It:

1. Place sealed jars in your water bath canner. Fill with water to cover the canner with 1-2 inches of water. Remove the lids, set the screw bands aside, and drop the sealing discs into the hot water, but not on top of the jars. Simmer at 180F as you prepare your jam. (This will sterilize the jars.)
2. Blanch tomatoes. When done, peel them and remove the seeds. Coarsely chop them. Measure 7 cups and set aside.
3. Finely chop and remove seeds from jalapenos. Set aside.
4. In a large saucepan, combine tomatoes, green pepper, onions, jalapenos, garlic, vinegar, tomato paste, and cilantro. Bring to a gentle boil. Stir occasionally—Cook for about 30 minutes, or until salsa reaches desired doneness.
5. Remove jars from the canner and dry them, taking care not to burn yourself. Dry the sealing discs, too. Ladle hot salsa into the jars, leaving ½" headspace. Use a nonmetallic utensil to remove air bubbles and add more salsa if needed. Wipe jar rim, center the sealing disc, and screw on a screw band until it is fingertip tight. Return the jar to the canner.
6. When the canner is full, cover all jars with at least 1" of water. Add the lid and bring the water to a full rolling boil. Boil for 20 minutes for altitudes up to 1000 feet.
7. Without tilting them, remove jars and place them on two-layered tea towels on the counter and leave them for 24 hours. Do not tighten screw bands. Check the

seals in 24 hours. If any are not sealed, transfer them to the fridge and use them first. Dry off all jars and loosely replace screw bands.

Sliced Peaches

Makes 4 x 500mL Jars or 8 x 250mL Jars

What You Need:

- 9 Large Peaches, sliced
- 1.5 cups Granulated Sugar
- 8 cups Water

How to Make It:

1. Place four sealed jars in your water bath canner. Fill with water to cover the canner with 1-2 inches of water. Remove the lids, set the screw bands aside, and drop the sealing discs into the hot water, but not on top of the jars. Simmer at 180F as you prepare your jam. (This will sterilize the jars.)
2. Cut a shallow "X" into the bottom of each peach. Boil a large pot of water and fill a large bowl with ice water. Boil the peaches for 3 minutes, then remove them straight to the ice water. When peaches are able to be handled, peel them.
3. Slice the peaches and discard the pit and skin. Divide the sliced peaches between the jars.
4. In a medium pot over medium heat, combine sugar and 8 cups of water. Simmer and stir until sugar has completely dissolved.

5. Remove jars from the water bath canner and seal discs. Wipe them dry without burning yourself. Pour syrup over peaches, leaving ¼" headspace. Gently tap jars to settle the peaches and syrup. Add more syrup if it is necessary to reach the ¼" headspace.
6. Return jars to water bath canner, making sure there's 1" of water over the jars. Bring the water to a boil and let boil for 25 minutes for altitudes up to 1,000 feet.
7. Remove jars from pot and let them sit on two-layered tea towels on the counter for 24 hours. Do not tighten screw bands. After 24 hours, check the seals. Any that failed to seal should be kept in the fridge and eaten first. Dry jars and screw bands and loosely return screw bands to sealed jars and store.

Pressure Canning

Pressure canning is a method of preserving that uses a large pot with a lid that is tightly sealed on top of it. Pressure canners reach far higher temperatures than simply boiling water in a pot. This is the only way to reach such temperatures with water effectively. These temperatures are required to kill *Clostridium Botulinum* so that the foods inside of the jars are safe. In foods that are not as acidic, such as meats and vegetables, this is necessary as it is more challenging to kill *Clostridium Botulinum* off of these foods than it is off of those that are more acidic, such as fruits.

What Foods Does Pressure Canning Work for?

Vegetables, fruits, meats, poultry, seafood, soups, and broth can all be safely preserved using a pressure canner.

What Materials Do You Need for Pressure Canning?

A pressure canner will need to be set up with the lid, the rack, and about 2-3 inches of water inside of it.

How Does the Pressure Canning Process Work?

Cans are placed on the rack within the canner, and the pot is sealed before it is heated to 240F. It will need to be maintained at that temperature for a period of time, based on what you are canning and what altitude you are at. When that time is reached, you will use the vent to release some of the pressure from inside of the can. Then you will carefully remove the lid, remove the jars using proper jar lifters. Then, you will let the cans sit overnight, and you will check them the next day. Do not tighten any of the lids. Any jars that did not properly seal should be consumed right away, while everything else can be stored for 1-2 years. Be very careful with pressure canners as they get much hotter than boiling water and can cause disastrous injuries. Further, the pressure can blow the lid off and cause serious damage to your house if you are not careful.

Note that with canning recipes, jars can be smaller than defined in the recipe but not larger. The amount of time that jar spends in the pressure canner will remain the same unless otherwise noted.

Canned Ground Meat

As Many As You Want. Pints or Quarts.

What You Need:

- Ground Meat

- Boiling Broth Or Water
- 1 TSP Salt Per Pound Of Ground Meat

How to Make It:

1. Sauté meat until just done. Add 1 TSP of salt per pound of meat you are using. Drain away any excess fat.
2. Fill clean jars with meat. Cover them in boiling broth or boiling water. Leave 1" headspace. Center sealing discs and screw bands until fingertip tight.
3. Process at 11 pounds of pressure in a dial gauge canner for 75 minutes if you are canning pints or 90 minutes if you are canning quarts if you are up to 2,000 feet altitude. Process at 10 pounds of pressure in a weighted gauge canner for 75 minutes if you are canning pints or 90 minutes if you are canning quarts if you are up to 1,000 feet altitude. Adjust pressure and cook time based on your altitude.

Canned Green Beans

As Many As You Want, Pints Or Quarts.

What You Need:

- Beans
- Boiling Water

How to Make It:

1. Wash young beans thoroughly. Chop off the stem, blossom ends, and "strings" if any appear. Leave whole or cut into 1" pieces. Make sure whole beans will fit into jars while being able to leave 1" headspace.
2. Pack beans tightly in clean, hot mason jars, leaving 1" headspace. Cover with boiling water, still leaving 1" headspace.
3. Process at 11 pounds of pressure in a dial gauge canner for 20 minutes for pints and 25 minutes for quarts if processing at altitudes of up to 2,000 feet. Process at 10 pounds of pressure in a weighted gauge canner for 20 minutes for pints and 25 minutes for quarts. Adjust pressure and cook time based on your altitude.

Canned Broth

As Many As You Want, Pints Or Quarts.

What You Need:

- Bone Broth (make it in advance by simmering a whole poultry carcass or fresh trimmed beef bones in a stockpot. Simmer poultry bones for 30-45 minutes, simmer beef bones for 3-4 hours.)

How to Make It:

1. Prepare your broth by simmering a poultry carcass for 30-45 minutes or beef bones for 3-4 hours. If using beef bones, rinse them before placing them in the water to prepare your stock with. After recommended cook time, remove the bones and let your broth cool to room temperature. Skim off any fat and discard—strain broth to remove bits of bone. If there was meat on the bones, you

can remove it and add it back to the broth for flavor if you want. Reheat broth to boiling.

2. Fill clean jars with hot broth, leaving 1" headspace. Center sealing discs and screw on screw bands to fingertip tight. Process at 11 pounds of pressure in a dial gauge canner for 20 minutes for pints and 25 minutes for quarts at altitudes up to 2,000 feet. Process at 10 pounds of pressure in a weighted gauge canner for 20 minutes for pints and 25 minutes for quarts at altitudes up to 1,000 feet. Adjust your pressure and times based on your altitude.

Chapter 7

Dehydrating

Dehydrating is a form of food preservation that involves removing the moisture content from foods so that 5% or less moisture remains in the food itself. Food is dehydrated in proper dehydrators, which are designed to wick away moisture and dry the foods inside. While you can use the sun to dehydrate foods, it is not recommended as you cannot control the temperature or the rate at which food is being dehydrated. It can take anywhere from a few hours to an entire day to dehydrate food. Dehydrated food will have different textures based on what was dehydrated, and how thick it was sliced before dehydrating. Bananas, apples, and onions, for example, will dehydrate to the point where they are crispy. Slices of meat, apricots, and mangoes will retain a somewhat chewy texture, even after they have been dehydrated. All dehydrated foods will need to be properly stored in airtight containers afterward to prevent spoilage. Foods that retain closer to the 5% moisture allowance should be kept in vacuum-sealed bags until you are ready to use them to prevent spoilage.

What Foods Does Dehydrating Work for?

Most fruits, vegetables, and meats can be dehydrated. Dairy products like cheeses and eggs can be dehydrated, too. You can dehydrate foods with the intention of eating them dehydrated, of rehydrating them later, or of adding them to baked goods later on. Certain foods, like eggs, will need to be cooked even if they have been dehydrated as they still run the risk of causing illness if you eat them raw.

What Materials Do You Need for Dehydrating?

To dehydrate, you are going to need a handful of materials. These materials will prepare the foods for the dehydrator, allow you to dehydrate the foods, and allow you to store the foods after they have been dehydrated.

They include:

- An electric dehydrator (get the largest one you can afford without compromising on quality)
- Dehydrator tray liners
- Cutting board
- Sharp knife
- Vacuum sealer with bags
- Blender

How Does the Dehydrating Process Work?

The dehydrating process works by first preparing foods to go into the dehydrator. In order to work efficiently, foods need to be sliced into thin slices so that they can dehydrate relatively quickly. If a food is too thick upon going into the dehydrator, you will end up either waiting excessive amounts of time for the food to dehydrate or finding yourself unable to completely dehydrate the foods to less than 5% moisture, which means bacteria will grow. Once your food is sliced, you will place it on the dehydrator trays, turn on your dehydrator, and dehydrate your foods until they have less than 5% moisture remaining. The amount of time required will vary depending on what you are dehydrating, so it is best to start all recipes in the morning to avoid having to check on your food into the later hours of the night and early hours of the morning.

If you are making something like dehydrated eggs, you are going to want to use a dehydrator tray liner, which is used to prevent wet ingredients from dropping through the tray. This way, they dehydrate correctly. Dehydrator tray liners can also be used to make fruit leathers. To make fruit leathers, you will blend up a fruit recipe in the blender until it is completely smooth and then pour it over the dehydrator tray liner before dehydrating it. Once it's done, you will have delicious fruit leathers.

Dehydrating can be done in an oven, too, although it will use up a lot more energy, and the foods may lightly cook in the process. While this does work, it is ultimately best to use a proper dehydrator, which can keep the right temperatures and conditions inside of the appliance for proper, safe dehydrating.

Dehydrated Potato Flakes

1 x 1 Pint Jar

What You Need:

- Five potatoes, peeled and chopped

How To Make It:

1. Add peeled and chopped potatoes to a pot and cover them with water. Boil the potatoes for about 10-15 minutes until they are fork-tender and ready to be mashed.

2. Move soft potatoes into a bowl, taking care not to add any water to the bowl in the process. Mash the potatoes until they are smooth. Refrain from adding anything to flavor the potatoes or soften the potatoes at this point, such as herbs or milk.
3. Lay potatoes on a fruit roll sheet on your dehydrator tray and dehydrate on 145F for 6 hours.
4. Break potato sheet into chunks, then blend them in your blender until they are ground into flakes. If you make your flakes too fine, they will be sticky when you cook them, so refrain from blending them too much.
5. Store potato flakes in a clean, airtight container in a cool, dry spot for up to 6 months.

Dehydrated Mango Slices

About 1-2 pounds of dehydrated mango.

What You Need:

- 5 ripe mangoes, peeled and pits removed
- ¼ cup lemon juice
- 1 TBSP raw honey

How to Make It:

1. Slice mangoes into thin strips.
2. Stir honey and lemon juice together. Dip thin mango strips into the mixture, shake off excess juice, and place them on dehydrator sheets.

3. Dehydrate for 10-12 hours at 135F. Around 8 to 9 hours, begin checking your mangoes for doneness. The exact time it will take depends on how ripe your mangoes were, how humid it is in your home, and how thin you sliced the strips.

Tasty Fruit Roll-Ups

Approximately 8, depending on what length you choose to cut yours.

What You Need:

- 5 cups of fruit, any fruit you prefer (you can also add some vegetables for added nutrition)
- ¼ cup raw honey

How to Make It:

1. Clean your fruits and vegetables for processing. Remove seeds and stems. The peels are nutritious, so you can keep them on for your fruit roll-ups; however, you can remove them if you decide you do not like the taste or texture. For foods where peels are not typically eaten, such as bananas, oranges, or pineapples, remove the peel.
2. Place all of your fresh fruits and vegetables into a blender with honey and blend until it is completely smooth. If you prefer even smoother textures, you can use a juicer to prepare your fruits and vegetables. However, there will be a lot of waste, and much of the pulp contains a great deal of nutrition, so it is better to blend your foods.

3. If you need the recipe to dehydrate faster, you can heat it up in a pot over medium heat for 10-15 minutes, stirring occasionally.
4. Spread your mixture out onto dehydrator trays that have been lined with tray liners or parchment paper. Thicken your puree around the edges of the tray, as they will dry faster. You should have about ¼" thickness at the edges of your trays and 1/8" thickness in the centers.
5. Dry your fruit roll-ups for 6 to 8 hours at 145F.
6. Roll your dried fruit roll-ups tightly and then cut them with a sharp knife. Wrap each piece with saran wrap and store it in an airtight container for up to 6 months.

Chapter 8

Freezing

Freezing is exactly what it sounds like: the type of food preservation that you engage in when you freeze foods. Virtually every household with a standard fridge has a freezer, which means you likely already engage in freezing foods at this time. Despite how common freezers and frozen food are, many people do not know how to use their freezers for the average ingredient. Instead, your freezer may typically be full of ingredients that you already bought frozen from the store, and you simply transferred them into your freezer.

Despite what you may think, not all things can just be tossed into the freezer. While most things are safe to be frozen, preparation methods must be followed to prepare those items to be frozen properly. Not following the proper freezing methods can lead to freezer burn, which can, in turn, reduce the quality of the food and drastically deteriorate the taste of the food, too. Each food type will have its own means for properly freezing said food.

What Foods Does Freezing Work for?

Most foods can be frozen, so it is actually easier to discuss what *can't* be frozen. Fresh tomatoes should never be frozen, though they can be cooked into sauces or pastes and then frozen that way. Whole eggs, rice, pasta, fried foods, salad greens, herbs, most sauces, cucumbers, and potatoes should never be frozen. While dairy products such as sour cream, yogurt, and milk may be frozen, know that the taste and texture will drastically change, so you will have to use them inside of a recipe if you are going to freeze them. Previously frozen meat should also not be refrozen. Although it will be safe to eat, it will seriously destroy the taste and texture of the meat.

What Materials Do You Need for Freezing?

Which materials you need for freezing depends on what it is that you are freezing. The following list of materials will provide you with everything you need to freeze all different types of foods, no matter what they are.

They include:

- Heavy-duty freezer bags
- Freezer-safe containers
- A large pot for boiling water
- A large bowl of ice water
- Tea towels for drying
- A pizza tray with holes in it
- A cutting board
- A sharp knife

How Does the Freezing Process Work?

The freezing process changes depending on what it is that you are freezing. When you are freezing things like meat or dairy products, you can freeze them by simply placing them in heavy-duty freezer bags or freezer-safe containers and freezing them that way. The same goes for most soups and sauces.

If you are going to be freezing fruits or vegetables with high water content, such as peppers, you are going to want to flash freeze them. You can flash freeze things by cutting them into the size you will want them to be in when you are consuming or using them later. Then, you will lay them on a pizza tray in a single layer, taking care not to let them touch each other too much. Pop the pizza tray into the freezer and let the ingredients freeze for 1-2 hours, or until they are completely frozen through. Then, remove them from the tray and place them in a heavy-duty freezer bag or a freezer-safe container. This way, they do not freeze together in one big lump.

Other types of vegetables, such as carrots, broccoli, or cauliflower, should be blanched before freezing. To blanch vegetables, you will boil a large pot of water, while also preparing an ice bath for the vegetables you are blanching. You will then cut the vegetables up into the sizes you want to freeze them in and place them in the boiling water for up to 1 minute. Choose a few as 15-20 seconds for vegetables that are already softer, such as broccoli or cauliflower. Immediately transfer the vegetables to the ice bath. Once they are completely cool, transfer them to a tea towel to dry. Then, follow the flash freezing method to ensure that they freeze separately and not in one big lump. When they are done, transfer them into a heavy-duty freezer bag or a freezer-safe container.

Frozen Green Beans, Summer Squash, and Tomatoes

Makes as much as you would like.

What You Need:

- Green beans, ends removed
- Summer squash, peeled and chopped

- Tomatoes

How to Make It:

1. First, prepare your green beans. Chop off the stems on the ends, then cut your beans into whatever lengths you want. Lay a piece of wax paper over a cookie sheet and spread the beans out in a single layer. Place the entire cookie sheet into your freezer and freeze the beans for 30 to 45 minutes, or until they feel frozen to the touch. Add them to your frozen mix bag.
2. Next, prepare your summer squash. To do this, you want to clean your squash first. Then, you want to slice it into rounds. If you want, you can also slice the rounds into quarters to create smaller pieces. Boil a pot of water and prepare an ice bath in a separate container. Once the water is boiling, add the squash to the boiling water and leave them there for three minutes. Begin counting as soon as you have placed your squash in the pot. When the three minutes are up, transfer them to the ice bath for five minutes. Then, drain them, dry them, and lay them on your flash-freezing tray and freeze them for about 30-45 minutes. When they are done, add them to your frozen mix bag.
3. Finally, prepare your tomatoes. To do this, wash and dry your tomatoes. Remove the stem and the core, then cut them into whatever size you desire. Place them on the flash freezing tray, skin side down. Freeze them for 30-45 minutes, or until they are frozen to the touch. Add your tomatoes to your frozen mix bag.
4. As long as you flash freeze all of your ingredients before adding them to your storage bag, they will not stick together too badly when you are ready to remove them from the bag to cook them. Be sure to keep them in a heavy-duty freezer-

rated bag or container to avoid having your vegetables sustain freezer burn, as this will deteriorate the quality of your vegetables by damaging both their taste and their texture.

5. Eat your vegetables within 6-12 months for the best quality.

Frozen Meat

Makes as much as you would like.

What You Need:

- Meat
- Plastic wrap
- Aluminum foil
- Freezer-friendly zip-top bag

How to Make It:

1. Keep your meat in the package in which you purchased it. Tightly wrap the entire package in plastic wrap, then in aluminum foil. Place the entire thing into a freezer-friendly zip-top bag and store it in your freezer for up to three months. After 3-6 months, the meat will still be edible. However, the taste will begin to deteriorate rapidly.

Frozen Bread

1 loaf, or as many as you would like.

What You Need:

- A loaf of bread
- Plastic wrap
- Aluminum foil
- Freezer-friendly zip-top bag

How to Make It:

1. Remove your bread from its original bag. Or, if you made your own bread, keep it unpackaged. Cover the loaf in plastic wrap, ensuring that it hugs the edges of the bread tightly so that no air can get to it. Wrap it in one more layer of plastic wrap. Wrap the entire thing in aluminum foil, then place it into a large freezer-friendly zip-top bag. If you do not have a large enough bag, wrap it in a second layer of aluminum foil. Freeze the bread for up to two months.

Chapter 9

Brining and Salting

Brining and salting are actually two different methods of the same preservation style. When brining, you use a wet brine to cure your meats, while salting means you use a dry brine to cure your meats. Wet brines are made of salt, sugar, and water and are accomplished by submerging the food item of choice in the brine with a heavy crock over the food item to keep it submerged in the brine. Dry brines are made of salt alone. They work by completely dusting your food in salt, which draws out the moisture and cures the meat. The trouble with dry brining, or salting, is that you might not be able to fully cover the meat in salt, which results in some of it spoiling. Further, if you add any additional salt, your food quality will be drastically reduced. For this reason, brining is the preferred method over salting these days, although both are still important to know about.

What Foods Does Brining and Salting Work for?

Brining and salting are mostly used with meats and seafood. Dry brining has also been used for vegetables like cabbage and runner beans.

What Materials Do You Need for Brining and Salting?

The tools you need to brine or salt food depend on if you want to brine or salt your food. Obviously, right? While not many tools are needed, it is important that you have the right ones to keep your food properly protected.

To brine your food with a wet brine, you will need:

- A brining bucket

- Brining salt
- A meat needle/pump
- A heavy crock

To salt your food with a dry brine, you will need:

- ½ TSP Kosher salt per pound of meat
- Fridge

How Does the Brining and Salting Process Work?

The brining process, or wet brining, works by combining brining sodium with water and pouring it in a brining bucket. Some brines will also call for a small amount of sugar to be added to the brine before adding it into the bucket. If you are in the wilderness or near the ocean, ocean water can be used for brining. However, you will need to bring it to a full rolling boil for about 5 minutes and then let it cool completely to kill off any unwanted bacteria and parasites.

Once the brine is prepared, you will pour it into the brining bucket. Then, you will take your meat needle or pump and draw brine up into the meat needle or pump, then inject it into the meat itself. Next, you will place the meat into the brining bucket and use a heavy crock to weight it down. You need to make sure the meat is completely submerged into the brine so that it is properly preserved. If any of the meat is above the surface of the brine, it will spoil.

For salting, or dry brining, you will rub the salt into the surface of the meat. Massage it until the meat is completely covered, and the salt sticks easily and evenly across the surface of the meat. You will need to leave it for 2-4 hours for a thinner cut or 2-4 hours for a larger one. Particularly thick cuts, such as poultry breasts, should cure for 4-6 hours. Roasts should cure for 12-48 hours. They can be kept in the fridge until you are ready to use them.

Brined Pork

4 Pork Chops

What You Need:

- 3 cups water
- ¼ cup firmly packed brown sugar (golden)
- ¼ cup pickling salt
- 2 cups ice cubes
- 4 pork chops

How to Make It:

1. Dissolve water, salt, and sugar. If you need to, you can do this over low heat on the stove, so everything dissolves faster. Add ice to your brine. Let the brine cool to lower than 45F.
2. In a zip-top bag, place 4 pork chops and cover them with brine. Place in the fridge for 2-6 hours, then use them in your desired recipe. If you are not ready to use them, remove them from the brine and store them for up to 2 days in plastic wrap in your fridge.

Brined Cheese

1 block of cheese.

What You Need:

- 2.25lb salt
- 1-gallon water
- 1 TBSP calcium chloride (30% solution)
- 1 TSP white vinegar

How to Make It:

1. Combine brine ingredients.
2. Use the freshest cheese you have, or make your own. Ensure it is cooled before placing it into your brine. Salt the top surface of the cheese, as the cheese will float, and the top surface will not be properly coated. Halfway through the brining process, flip the cheese and salt the top once again to protect it. For hard cheeses, you will need to brine them for 24 hours. For softer cheeses, they will require 12 hours of brine.
3. After the required amount of time, remove your cheese from the brine and allow it to air dry. It should take 1-3 days for the air-drying process. You will know that it is done when you can see a firm, dry surface on your cheese. At this point, you can wax your cheese or allow it to develop a natural rind. If you allow it to develop a natural rind, be sure to find instructions on how to do so with the specific cheese you are using, to avoid accidentally causing spoilage in your cheese.

Chapter 10

Sugaring

Sugaring is similar to pickling; however, it works a little differently. To sugar something, you will first dehydrate it; then, you will pack it with sugar. The sugar creates a hostile environment for bacteria, which prevents spoilage. Sugaring can be used in a pinch, and for some things, it is great. However, you must be cautious with this method as sugar itself can attract moisture, which, as you know, can spoil foods. If moisture rises in the sugar itself, native yeast in the sugar and the environment will come out of dormancy and begin to ferment the sugars. While fermentation is a form of food preservation, when it is accomplished by accident, it can lead to unpleasant situations where the food itself would taste gross. It could lead to you falling ill if you consume it. Following exact methods when sugaring food and storing sugared food is the best way to make sure you are not putting yourself at risk of getting sick from sugared foods.

What Foods Does Sugaring Work for?

Sugaring is usually used with fruits; however, certain vegetables such as ginger can be stored using the sugaring process.

What Materials Do You Need for Sugaring?

Sugaring food is done by completely dehydrating foods, than by cooking them in liquid sugar. You then remove the item from the sugar syrup and allow them to dry completely before storing them in airtight containers.

To complete the sugaring process properly, you will need:

- Fully dehydrated food
- A large pot
- Sugar
- Water
- Airtight containers or a vacuum sealer

How Does the Sugaring Process Work?

The sugaring process works using any raw sugar or table sugar that is in crystallized form. However, syrup, honey, and molasses can all be used to store foods this way, too. Before you begin the preservation process of a fruit or vegetable, you will thoroughly wash it and then dehydrate it completely. Then, you will cook the dehydrated food in liquid sugar until the food itself becomes crystallized. Once it has, you can remove the food from the sugar, dry it, and preserve it in airtight containers or bags until you are ready to use them. Because of how finicky sugared food can be, a vacuum sealer is generally the best way to preserve your sugared foods as it will ensure that no moisture can get into your final product.

Sugar Preserved Citrus Fruits

1 Pint Jar.

What You Need:

- Citrus (make a mix of your favorite citrus fruits such as oranges, limes, lemons, etc.)
- Granulated sugar

How to Make It:

1. Quarter your fruit and place a layer into a sterilized canning jar. Sprinkle with one tablespoon of sugar. Repeat this until the entire jar is full. Then press down on the fruits, which will release some juices into the bottom of the jar.
2. Cover the entire contents of the jar in 2-3 more tablespoons of sugar, as much as you can fit.
3. Place the jar in the fridge for three weeks so it can "rest."
4. Eat your sugared fruits anywhere from 3 weeks to up to 6 months.

Chapter 11

Smoking

Smoking is a type of food preservation method that results in foods tasting similar to barbecued dishes. The difference with smoking is that it has a delicious, naturally smoky flavor based on the type of preservation method used and that it can be preserved for much longer, and much easier than a barbecued dish can be. While a barbecued dish must be stored in the fridge and eaten within a few days, smoked foods can be vacuum sealed and eaten within a few months or up to a year.

What Foods Does Smoking Work for?

Meats and fish are the most common foods to be smoked, though you can also smoke cheese and certain vegetables. Some ingredients that are used in beverage making are smoked, as well, to give beverages a nice smoky flavor. Since you are preserving food for your family, it is most likely that you will be smoking different meats and fish.

What Materials Do You Need for Smoking?

In order to start smoking your foods, there are a few different tools you are going to need. A smoker, fuel, and wood chips are the first tools you will need. You will also need utensils for placing and removing your smoked preserves, and meat or fish for you to smoke.

Types of Grills

There are two types of grills you can use when it comes to smoking meats. Your classic outdoor barbecue grill is one you can use, or you can purchase a proper smoker. If you are going to be smoking a lot of meats, it is better to buy a smoker as it will do a much better job of smoking the foods you will be preparing.

If you are using a grill, you are going to need to be able to create enough smoke to cook your meats effectively.

On a charcoal grill, you can do this by placing coals on one side of the barbecue and placing a drip pan on the other side. Once your coals are hot enough to cook over, place a layer of wood pellets over them. If you place a layer of liquid in your drip pan, you will get much better smoking results this way. Apple juice is a great liquid to use here as it will add flavor to your meat during the smoking process. When you are ready to start cooking, place your meat over the drip pan. Be sure to keep a small opening in your lid so that the smoke can vent out properly. This keeps smoke moving around the meat so that it can properly penetrate the surface and cook it through.

If you are using a gas grill, there are no charcoals for you to lay the wood over. In this case, you need to place your wood chips in a metal pan, which will be inserted directly over the gas flames. Set this pan to one side. Let the entire setup preheat for about twenty minutes before placing your meat on the side opposite where the wood chips were wet up. Then, close the lid leaving space for ventilation to occur during the smoking process.

Types of Wood

There are many different types of wood that can be used with smoking recipes. The type of wood you choose will affect the flavor of your meat, so you want to pick one that is going to give you the best flavor. Avoid using softwoods like pine or cedar, as these will produce too much smoke and will release a resin that will penetrate your meat and create a resinous flavor.

The types of wood that are best to use with smoking, and what types of meat they can be used for, can be found here:

- Alder: Chicken, seafood, pork.
- Apple: Chicken, seafood.
- Cherry: Chicken, seafood, pork, beef.
- Hickory: Pork, beef.
- Maple: Chicken.
- Mesquite: Pork, beef.
- Mulberry: Chicken, seafood, pork.
- Oak: Chicken, seafood, pork, beef.
- Peach: Chicken, pork.
- Pear: Chicken, pork.
- Pecan: Chicken, pork, beef.
- Walnut: Pork, beef.

While you can certainly use any of these woods with any meat you like, they are unlikely to taste good with meats that are not listed here. Chicken and seafood tend to pair well with sweeter, lighter woods since they absorb more flavor, which means heavier woods can become too strong in these meats. Alternatively, pork and beef do not absorb quite as much of the flavor, which means that you can go ahead with stronger and heavier woods because they will penetrate the meat and leave a better flavor in the end.

When you purchase your meats, you will notice that there are many options. Wood chunks, chips, and pellets are all offered as wood sources for smokers. For longevity and sustainability, chunks and chips are the only types of wood you should really consider. While pellets will work, they tend to burn off quickly, which will leave you having to add more far more frequently than you would with chunks or chips.

Wood chunks are excellent if you are going to be smoking food all day long. If you are going to be preserving large amounts of food with your smoker, these chunks will save you from having to add more wood to the smoker as often. This prevents you from opening the smoker too frequently, meaning your food cooks better. To get even more smoke out of your wood chunks, you can soak them in water for up to one hour before you begin smoking your food so that they are nice and moist, as this will cause them to smoke more.

Wood chips are great if you are only going to be smoking food for a couple of hours at a time. These are smaller than the chunks, and so they will burn faster, though that is not a problem if you will not be using your smoker all day. The best way to get the most out of your chips is to soak them in water for up to 30 minutes before you start so that they produce more smoke.

How Does the Smoking Process Work?

Smoking your food requires you to provide your food with indirect heat over lower cooking temperatures and with an abundance of smoke present. The smoke itself creates an acidic coating on the surface of the food, which prevents bacteria from growing on or in the meat. It also dehydrates the food, which creates an environment that is less hospitable for bacteria to grow. Smoking also enhances the flavor of the food you are preparing, making it an excellent option for preserving foods. Smoked foods can be served as is, or they can be placed in vacuum-sealed bags and stored for longer periods of time. The exact amount of time a food can be stored depends on what that food is and how well it was smoked.

There are two types of smoking: hot smoking and cold smoking. Hot smoking happens at temperatures over 150F, while cold smoking happens at temperatures less than 100F. Hot smoking is used for cooking the food while also providing a rich smoky flavor, while cold smoking will only flavor the food, it will not cook it. Certain already-cured meats such as salami and cheeses such as cheddar are cold smoked as a way to enhance their flavor.

The first step of any smoking recipe will always be to prepare your food first. For meats, this requires you to trim the meat into the desired shape and then marinade it in a dry brine, which will help cure the meat. You will store the meat with the dry brine in the fridge for up to five days, flipping it each day to ensure that the cure evenly distributes across the meat. Once five days have passed, you will rinse the meat under cold water and allow it to dry out in the fridge. The next day you will prepare your smoker by adding the necessary wood chips or chunks and allowing it to preheat for about 20 minutes, as this will ensure that it is nice and hot and the smoke is abundant for the meat. Then, you will place the meat in your smoker. Each type of meat will require different smoke times, so you will need to follow a specific recipe for your desired ingredient.

By following this entire process, you effectively prepare the meat for smoking by helping to first dry it out using the dry brine. Then, come the smoking day, you allow a combination of heat and smoke to properly cook the meat while also creating an acidic environment on the meat which prevents it from going bad. After the entire smoking process, the meat will be dehydrated enough that it will be able to be preserved for long periods of time. The key to success here, though, is to realize that the meat will not be completely free of moisture, which means bacteria *can* still grow on it. A proper vacuum-sealed bag will prevent the moisture content from increasing or spreading and will ensure that you can enjoy your meat for months to come.

Smoked Pastrami

Five to six pounds of pastrami, depending on the size of the meat you cook.

What You Need:

- 1 brisket (5-6 pounds)
- 1 cup + ½ cup brown sugar
- ¾ cup kosher salt
- 2 TBSP mustard seed, whole
- 2 TBSP paprika
- 1/3 cup + 2 TBSP coriander seed, whole
- 1/3 cup + 2 TBSP black peppercorns, whole
- 1 TBSP ground ginger
- 1 TBSP red pepper flakes
- 1 TBSP allspice berries, whole
- 1 TBSP granulated garlic (not garlic salt)
- 1 TBSP granulated onion (not onion salt)

- 2 TSP pink curing salt
- 6 cloves, whole
- 4 garlic cloves, whole
- 6 bay leaves
- 1 gallon filtered water

How to Make It:

1. Trim the fat off the top of the brisket.
2. Pour 1-gallon water into a large, non-reactive container. Add 1 cup brown sugar, ¾ cup kosher salt, 2 tbsp whole mustard seed, 2 tbsp whole coriander seed, 2 tbsp whole black peppercorns, 1 tbsp ground ginger, 1 tbsp allspice berries, 2 tsp pink curing salt, 6 whole cloves, 4 whole garlic cloves (peeled,) and 6 bay leaves.
3. Wrap the brine container and cure the brisket for 7-10 days, flipping it over daily to ensure that it cures evenly.
4. On cook day, preheat your smoker to 225-250F. Add your desired wood chips or chunks to the coals. Close your smoker and let the smoke begin to develop inside.
5. Prepare your dry rub by combining ½ cup brown sugar, 2 tbsp paprika, 1/3 cup whole coriander, 1/3 cup whole black peppercorns, 1 tbsp red pepper flakes, 1 tbsp granulated garlic, and 1 tbsp granulated onion.
6. Thoroughly rinse the brisket under cool water, then pat it so that it is completely dried. Add the dry rub and massage it into all sides of the brisket, applying liberally as you go.
7. Place your brisket in the smoker and let it smoke until it reaches 150F inside. This should take around 3-5 hours.

8. Preheat your oven to 250F. Pour water into a roasting pan so that it is 1" deep, and place a rack over it so that the pastrami is not sitting in water. Add an oven-safe probe into the pastrami, place it on the rack, and cover it with tinfoil. Cook until your pastrami reaches 230F inside. This should take around 4-5 hours.
9. Slice your pastrami and serve it, or store it in vacuum-sealed freezer bags for up to one year.

Maple Smoked Bacon

3 pounds of bacon.

What You Need:

- 3lb unsliced pork belly, about 1" thick and 6-8" long.
- 3 TBSP dark brown sugar
- 3 TSP black pepper, ground
- 3 TBSP pink curing salt
- 3 TBSP maple syrup

How to Make It:

1. Combine dark brown sugar, black pepper, pink curing salt, and maple syrup in a non-reactive container filled with about ½ gallon of filtered water. Add the pork belly to the container and let it sit for seven days, turning it daily for even curing.
2. After the 7th day, completely rinse the pork belly and let it sit in the fridge on a wire cooling rack for 12-24 hours, until it develops a sticky skin. This sticky skin is called "pellicle."

3. Prepare your wood pellets in your smoking machine and set the temperature to 165F. Let it preheat for around 20 minutes. Add your pork belly and smoke it for about 6 hours, or until it is 155F inside.
4. Slice the bacon. Store it in vacuum-sealed bags in the fridge for up to 2 weeks or in the freezer for up to 1 year. When you want to cook it, defrost it and fry it in a pan as usual.

Smoked Jerky, Black Pepper Flavor

4-6 Servings of smoked jerky.

What You Need:

- 1 bottle dark beer
- 1 cup soy sauce
- ¼ cup Worcestershire sauce
- 4 tbsp black pepper, ground and divided
- 3 tbsp brown sugar
- 1 tbsp pink curing salt
- ½ tsp garlic salt
- 2 pound trimmed beef top, sirloin tip, bottom round, flank steak, or wild game

How to Make It:

1. Make a marinade from beer, soy sauce, Worcestershire sauce, 2 tbsp black pepper, brown sugar, curing salt, and garlic salt.

2. Trim fat away from your meat and slice it into ¼" thick strips, against the grain. If it is too hard to slice your meat, place it in the freezer for about 30 minutes, and try again. This should firm up the meat, so it is easier for you to cut.
3. Place beef slices in a large zip-top bag and fill it with the marinade. Massage the bag to penetrate the meat with the marinade and to ensure that everything is evenly covered. Seal the bag and place it in the fridge overnight, or for up to 24 hours.
4. Prepare your smoker with a temperature of 180F. Remove beef slices from marinade and dry them between paper towels. Sprinkle the dried slices with the remaining 2 tbsp black pepper. Discard the remaining marinade.
5. Lay beef jerky strips in a single layer on the grill grate. Smoke for 4-5 hours, or until the jerky is chewy. It should be somewhat pliant when you attempt to bend a piece of your jerky.
6. When the jerky is done, transfer it to a zip-top bag while it is still warm, and let it rest on the counter for one hour at room temperature. Seal the bag, then place it in the fridge for up to 2 months. Or, vacuum seal the bags and place them in the freezer for up to 7-12 months.

Smoked Fish

One full fish, smoked.

What You Need:

- 5 pounds salmon, trout, or char
- Maple syrup

- 1 quart filtered water, cool
- 1 cup brown sugar
- 1/3 cup kosher salt

How to Make It:

1. Begin by curing your fish. Combine water, brown sugar, and salt in a non-reactive container and place the fish in it. Put it in the fridge for at least 4 hours, but up to 48 hours. Never go over 48 hours, or your fish will be too salty. Fish is softer than other meats, so it absorbs the liquid much faster. Flip the fish halfway through your curing time for even curing.
2. Remove the fish from the brine and pat it dry without rinsing the fish. Place the fish on a cooling rack, skin side down, and let them cool at 60F or cooler for 2-4 hours. Alternatively, you can cool it overnight in your fridge.
3. Use soil to lightly coat the skin of your fish, as this prevents it from sticking to the rack in your smoker. Then, place the fish on your smoker's grill, skin side down. Smoke your fish between 140F and 150F for an hour, and then finish it at 175F for 1-2 hours, or until the fish is cooked through.
4. Every hour that your fish is in the smoker baste it with maple syrup. This will remove any albumin that may form so that your fish stays tasty and fresh. Avoid letting the heat get too high, as this will result in more albumin and can overcook your fish.
5. Once the fish is done, cook it to room temperature for one hour, then wrap it in an airtight container and place it in the fridge for up to 10 days. Or, you can vacuum seal your fish and store it in the freezer for up to 1 year.

Chapter 12

Pickling and Fermenting

Pickling and fermenting are two types of preservation methods that use salt brine to create a delicious, sour outcome. Pickles themselves are something we are likely all familiar with; however, you may not be familiar with fermented foods and the taste of fermented ingredients. A great example of fermented foods that you can buy at the grocery store includes sauerkraut, horseradish, and kimchi. When you ferment food, you create an environment for that food through the help of salt brine and temperature that allows it to begin the fermentation process in a controlled manner. The result of properly fermented foods is the creation of food that is rich in probiotics, and that promotes healthy gut flora.

Fermentation and pickling may sound similar, but there are differences. With pickling, you are soaking food in acidic liquids to achieve a sour flavor, and preserving that food happens by proxy. Most pickled foods are pickled using a vinegar brine. With fermented foods, the sour flavor you taste is the result of a chemical reaction that occurs within the food between the natural sugars in the food and naturally present bacteria. You do not need acidic liquids to create a fermented food. Most fermented foods are fermented using a salt brine. You can make fermented pickles, as well, which is a unique recipe that combines the properties of both of these preservation methods to create a delicious result.

What Foods Does Pickling Work for?

There are many different types of food that you can pickle. The most obvious is cucumber, though you can also pickle asparagus, beets, bell peppers, cauliflower, carrots, green beans, onions, peaches, radishes, turnips, and even certain fish such as herring.

What Materials Do You Need for Pickling?

To properly pickle any food, all you need is pickling salt, vinegar, glass jars with proper sealing lids, vegetables or fruits, and herbs. You will also need a nice dark place to store your pickles as they absorb the liquids, *or* you can store them in your fridge if you are making fridge pickles. Fridge pickles will take up more space in your fridge. However, they will last much longer than fresh vegetables being stored in your fridge, so this does technically preserve them for a more significant period of time.

How Does the Pickling Process Work?

Pickling with vinegar brine works by using a simple mixture of water, salt, and vinegar to create a brine that effectively pickles your foods. There are many sub-methods of pickling, including pickling methods that allow for different flavors and that work for different types of foods. The distinction between these methods generally includes using different herbs or different ratios in the vinegar brine to create your desired flavor. Pickles can either be water bath canned so that they can be preserved at room temperature, or they can be made in the fridge, which means they are not sealed or shelf-stable, but they do last much longer than fresh vegetables would. Fridge pickles tend to mature faster than shelf-stable pickles, so they both have their own set of benefits.

As a food item sits in the vinegar brine, it absorbs the brine, which gives it a nice, sour flavor. This flavor will directly reflect the brine you used and anything that was placed in the brine with your pickles. This is how you make the flavors present in garlic pickles, dill pickles, spicy pickles, and other flavors that you might come across.

Cucumber Carrot Fridge Pickles Recipe

2 x 250mL Jars of Pickles

What You Need:

- 3 cups water
- ¼ cup rice wine vinegar
- 2 TBSP sugar
- 1 tbsp + 1 tsp kosher salt
- 1 cucumber, thinly sliced
- 2 shallots, thinly sliced and separated into rings
- 1 carrot, thinly sliced
- 1 red Thai chile, stem removed and thinly sliced

How to Make It:

1. Boil 3 cups of water. In a bowl, combine cucumber, shallots, and carrot. Cover the produce in 1 tbsp salt. Ensure the produce is in a heat-proof bowl. Pour boiling water over the picture. Stir, and let stand for 20 minutes.
2. Drain vegetables in a colander until they are completely drained. Combine vinegar, sugar, chile, and 1 tsp salt. Dissolve the sugar and salt completely. Divide

between 2 x 250mL jars, and then divide the produce between the jars, too. Place in the fridge for at least 1 hour. The longer they sit, the better they will taste. They can be kept for up to 1 year.

Pickled Egg Recipe

12 eggs.

What You Need:

- 12 eggs
- 1 cup white vinegar
- ½ cup filtered water
- 2 TBSP coarse pickling salt
- 2 TBSP pickling spice
- 1 onion, thinly sliced
- 5 black peppercorns

How to Make It:

1. First, hard boil your eggs by placing them in a pot and covering them in cold water. Bring the pot to a boil and immediately remove the pot from the burner, letting it stand for 10-12 minutes. Cool the eggs in a cool bowl of water, then peel the shells off. Place them in a 1-quart wide-mouthed jar, or divide them between two half quart wide-mouthed jars.
2. Combine vinegar, water, salt, and pickling spice in a saucepan. Add most of the onion, except for a few slices, and the five black peppercorns. Boil the mixture

until it reaches a rolling boil, then pour the mixture over the eggs. Add the reserved onion slices on top and cool to room temperature. Place in the fridge for three days before serving and eat within a year. Or, use a water bath canner to preserve the eggs for a longer period of time.

What Foods Does Fermenting Work for?

Fermenting foods is most common for vegetables and fruits. The fermentation process can also be used to make wine, as well as wild-caught yeast in the form of a sourdough starter. Fermentation happens when you allow the natural sugars present in an ingredient to engage in the fermentation process in an intentional, controlled environment. The best foods to ferment include, cabbage, carrots, cauliflower, cucumber, garlic, kohlrabi, pepper, radish, snap beans, and turnips. Foods with high water content work best for fermentation.

What Materials Do You Need for Fermenting?

Fermenting requires you to have a fermenting crock or jar, vegetables, herbs, pickling salt, plastic bags, and water. You will also need a sharp knife to fine-cut your fermented foods, as thicker foods do not ferment as well as thin-sliced foods.

How Does the Fermenting Process Work?

The fermenting process works by taking salt and massaging it into the vegetable of choice until it begins to wilt, and water begins to be extracted from the vegetable. It should take around three to five minutes to massage your vegetables to the point where they are ready to ferment. Once you have effectively massaged them, you will stuff them firmly, but not too tightly, into mason jars. If you have proper food-grade pickling crock, you can do this directly in the crock and then simply leave the ferment in the crock. If you do not have enough liquid to completely cover the food that you are fermenting, you will need to add a saltwater brine to it. Then, you will place a weight over the food to ensure that it stays under the brine. Unless you have a proper crock set up, the easiest way to do this is to insert a (clean) plastic bag into your mason jar. Then, fill it with the remaining brine, or with fresh filtered water, and seal it. This should create a closed seal over the top of the food, keeping it submerged in water and protected from air. It is important that fermented foods are kept in air-tight containers to avoid having air access the fermenting food. Air can increase the likelihood of mold forming or spoilage occurring before the fermentation process has a chance to really get started. Each day you will stir the fermented food and then replace the water bag into the jar to create the airtight seal again. You will want to keep the jar on a plate or in a small bowl, as it will "burp" air through the sides of the bag, which will bring some liquid with it, too. Most fermented foods are finished anywhere from a week to four weeks, depending on what flavor you are looking for. Once you reach your desired flavor, you will place your ferment into sterilized mason jars, leaving ¼" headspace. You can then put it in the fridge or water bath can it, depending on what your unique recipe calls for. If you place your ferment into the fridge, the fermentation process will be extremely slowed down to the point where you will not likely notice it.

The only type of fermentation that does not include salt is a sourdough starter. Sourdough starter is technically a fermented food, though it does not require salt, and it cannot be eaten as is. With sourdough, you do a mixture of flour and water on your counter and continue to "feed" it with more flour and water every single day. This results in the starter capturing wild yeast, meaning you do not have to use active yeast to create your sourdough bread. This starter can then be incorporated into sourdough recipes such as bread, pancakes, pizza crust, pretzels, cinnamon buns, and more. It is said that sourdough is easier to digest because of the fermentation factor, which makes this a great healthy bread option and also makes for an excellent alternative to store-bought yeast. Give the effects of the 2020 pandemic, this is a wonderful alternative if you find that your local store is sold out of conventional yeast.

Traditional Fermented Sauerkraut

Makes about 1 to 1.5 quarts of sauerkraut.

What You Need:

- 1 medium head green cabbage
- 1.5 tbsp kosher salt

How to Make It:

1. Completely clean all of your preparation tools, first. This way, your ferment has the best chance at developing healthy, clean bacteria, rather than the bad kind that will cause spoilage.
2. Remove the outer leaves of cabbage. Quarter it and cut out the core. Slice wedges into thin ribbons of cabbage and place them in a large, non-reactive bowl. Sprinkle 1.5 tablespoons of salt over your cabbage and start massaging it with your clean hands. As you massage it, the cabbage will begin to wilt, and water will begin to pour out of the leaves. It will start to look like coleslaw, rather than raw cabbage.

3. Pack sauerkraut into clean 500mL mason jars. It should take about three mason jars, as you want to leave at least 1.5-2 inches of headspace in each jar. Pour the liquid released by the cabbage out of the bowl and into the jars. If the liquid does not cover the cabbage, combine 1 tsp of salt with 1 cup of water until the salt dissolves, then pour that into your jars to cover the cabbage.
4. Weigh the cabbage down either using a smaller jelly jar full of clean stones or marbles or using a bag full of excess brine or filtered water. If using a jar, cover the entire thing with a cloth and use a rubber band to secure it in place. If using a bag, place a clean bag inside of the jar over top of your sauerkraut and use your hand to lightly press the bag out from the inside of the bag itself. Then, fill it with brine until the brine reaches the mouth of the jar. Remove all air from the bag and seal it, then move it around a bit to release any air bubbles that may have become trapped under the bag.
5. Over the next 24 hours, press the cabbage down about 3-4 times so that it becomes even more wilted.
6. Let the sauerkraut ferment between about 65F and 75F for 3-10 days. At temperatures over 75F, your sauerkraut will ferment rapidly and may become fairly soft in the process. Below 65F, your sauerkraut will not have enough warmth to encourage the growth of the good bacteria in the ferment.
7. As your sauerkraut begins to come "mature," you can start tasting it. Early on, it will have a salty taste. Over time, it should develop a sour, tangy scent, and taste, which indicates that it is done. If you see scum starting to develop on the top of your sauerkraut, that is normal. Skim it off and discard it. If you see any mold, discard your entire jar and start over, as mold can make any cabbage that

touched the mold dangerous to eat, and it can damage the taste of the rest of your sauerkraut.

8. Your complete sauerkraut can be transferred into either 4 x 250mL jars or 2 x 500mL jars once it is complete. Then, store it in the fridge. It can be eaten for up to 2 months, though it will typically last much longer. So long as it smells good to eat, and tastes good to you, the sauerkraut should still be fine to eat. Ferments like sauerkraut will rarely spoil, especially when kept in the fridge; however, they can begin to taste off over time.

Fermented Beet Slaw Recipe

Makes about 1 quart.

What You Need:

- 6 cups raw beets, peeled and shredded
- 1 tbsp kosher salt

How to Make It:

1. Lay shredded beets in a mixing bowl and sprinkle the salt over them. Let them sit for about 5 minutes, then use your clean hands to massage the beets to ensure that they all absorb the salt. As you do this, the shredded beat pieces will become wilted and will expel even more water.
2. Pack the beets into a clean, quart-sized jar and leave at least 1 inch of headspace. This way, the beets can safely release juices and gasses during the fermentation process. You will need to "burp" the jar every so often to release the gasses so that

the jar does not explode. You should burp your jar 2-3 times a day, or more if you notice that it is becoming particularly concentrated.

3. Ensure that the lid of your slaw is airtight to avoid any air getting into your ferment.
4. After 2-3 days, the beet slaw will be done. Refrigerate it for up to 2 months.

Chapter 13

Ash, Oil, Honey

Ash, oil, and honey are all used as a means for preserving foods. Each one of them has its own unique benefits and capabilities. They also share the common qualities of being lesser-known preservation methods that have stood the test of time and offered great preservation solutions to people for many generations. With all three of these preservation methods, the main goal is to coat the food you are preserving so that no air can access it. They also have their own antibacterial properties, which enable the foods within them to remain safely stored and able to be consumed for long periods.

What Foods Does Ash Work for?

Ash is an incredibly traditional food preservation method that can be used to store things like eggs, cheese, and tomatoes.

What Materials Do You Need for Ash?

In order to effectively preserve your food in ash, you will need fresh, sifted ash, an airtight container, and the food you intend on preserving.

How Does the Ash Process Work?

The process of storing your food in ashes is incredibly simple. You will take sifted ash and place a layer on the bottom of an airtight container. Then, you will layer your food into the ash, and then you will cover it with ash again. You will continue forming these layers until the box is full. Then, you will replace the lid on the box and store it this way for up to three months.

Ash Preserved Gruyere Cheese

Makes 1 block of ash-preserved cheese.

What You Need:

- 1 block of Gruyere, not a thin piece
- 3" of sifted wood ash

How to Make It:

1. Remove Gruyere from the fridge and let it dry off at room temperature for an hour or two so that it is not retaining too much moisture.
2. Cover the bottom of a stoneware pot in 1.5" of sifted wood ash. Place Gruyere cheese over the ash and cover it in another 1.5" of sifted ash. Store the pot of cheese in a cool root cellar for up to 3 months.

Ash Preserved Garden Fresh Tomatoes

Makes 15 tomatoes.

What You Need:

- 15 tomatoes
- 3" or more of sifted wood ash

How to Make It:

1. Pick tomatoes fresh from your garden. Choose ones that are ripe but not soft or overripe. Do not store tomatoes with bruises or blemishes, as it will not work. Line a wooden or cardboard box with paper, then place 1.5" of sifted wood ash in the bottom of the box. Layer your tomatoes onto the ash, taking care not to let them touch each other. Cover them with more sifted wood ash until there is 1.5" of ash over the tomatoes. Place it in a cool, dry place, such as a root cellar, for up to 3 months.
2. This process will result in the skin of the tomatoes wrinkling somewhat over time. However, the inside will remain juicy and fresh the entire time.

What Foods Does Oil Work for?

Oil is best for preserving foods like sun-dried tomatoes, baby artichokes, sweet peppers, eggplants, mushrooms, garlic, goat cheese, basil, lemons, sardines, and tuna fish.

What Materials Do You Need for Oil?

To properly store food in oil, you will need a clean, airtight container, olive oil, and the food that will be stored in the oil.

How Does the Oil Process Work?

The process of storing food in oil is simple. You will place your chosen food into an airtight container and then cover it with oil. Then, you will place the lid on your food and allow it to sit. The oil prevents any air from accessing your food, which means it lasts longer. For certain foods, such as seafood, you will need to cure them with salt first so that they are ready to be preserved in oil. Otherwise, they will not be safe to eat, and they can contaminate the oil with unsafe materials, too.

Garlic Preserved in Olive Oil

About 15 cloves of garlic, or more as needed.

What You Need:

- 15 cloves of garlic, peeled
- 250mL of olive oil

How to Make It:

1. Place garlic in a 250mL mason jar. Pour oil into the jar until it has completely covered all of the garlic. Close the lid on the jar until it is finger tight. Place it in the fridge for up to 3 months. Replenish the olive oil as needed.

Fresh Tuna Preserved in Olive Oil

1lb tuna in oil.

What You Need:

- 1lb tuna
- 1.5 cups olive oil
- 1 garlic clove, crushed
- 3 sprigs thyme

How to Make It:

1. Rinse tuna thoroughly, then use a paper towel to pat it dry. Cut the tuna down into smaller 1" portions. Place the tuna in a small saucepan with 1.5 cups of olive oil.
2. Heat the tuna over low heat until small bubbles start to come to the surface. Let it cook for 10 minutes, without letting the oil come to a boil. You want it to maintain a nice, low simmer that will gently heat the tuna without deep frying it.
3. When the tuna is cooked through, add the garlic and thyme, then pour the contents into a one-quart mason jar and let it cool to room temperature. Cover the jar, place it in the fridge, and let it infuse overnight before serving. You can save tuna this way for up to 2 weeks in the fridge.

What Foods Does Honey Work for?

Honey works best in specific canning recipes that call for honey, or with dehydrated fruits or herbs.

What Materials Do You Need for Honey?

To store things with honey, you will need raw, unpasteurized honey, a container for storing everything in, and dehydrated fruits or herbs.

How Does the Honey Process Work?

Using honey to store your food is a great alternative to using sugar. Honey can be used to replace sugar in certain canning recipes. Though, food can also be stored directly in honey since honey itself is antibacterial, and it is so thick that it will prevent any oxygen from getting to the food that is being preserved. In fact, this is why honey often makes for a great dressing for fresh wounds.

Aside from using honey to substitute sugar in canning recipes, foods can be stored directly in honey as well. To do so, you will dehydrate, or at least partially dehydrate, fruits, and then place them in honey. The honey would then keep them from fermenting, rotting, or spoiling for several months.

Honey Berry Puree

Makes 1 pint of honey berry puree

What You Need:

- 500g raw honey
- 160g strawberries

How to Make It:

1. Clean strawberries and remove the stems. Puree them in a blender without adding any water.
2. Mix the puree and honey together in a bowl. Pour the mixture into a pint-sized mason jar and seal. The honey will be shelf-stable as is, without canning it or putting it in the fridge. You should eat it within three months for freshness.

Chapter 14

Food Preservation Safety FAQ and Tips

As you likely already know, you have to be exceptionally careful with how you handle your food. When it comes to food preservation, the methods for keeping your food safe are slightly altered. In typical cooking, you use fresh food and heat it to specific temperatures to ensure safety. You may be familiar with storing your food in the fridge and in the freezer. However, these are two preservation methods that are generally easy and commonly accepted in our modern society. That means you have a lot of experience with safely storing your food in the fridge and freezer and preparing meals from your refrigerator and freezer.

When it comes to other methods of preserving food, especially methods that will be storing food at room temperature for any period, there are many things that you need to consider. These things will allow you to ensure that your food is preserved in optimal conditions so that you are confident that it is safe for you to consume those foods.

Understand that of all of the food preservation methods I have provided you within this book; each will offer you different lengths of preservation. Some will only extend the life of your food by a few weeks, while others will extend it by months or even years. This is just one factor that you need to be aware of when it comes to properly preserving your food so that you can create a hearty stockpile for your family. It is crucial that you consider best by dates and food spoilage in your plans when it comes to preserving foods, as you want to build a preservation schedule that will provide you with plenty to eat. Some things you may have to preserve multiple times throughout the year, while others may be safe to preserve only once per year or even once every other year. Naturally, the food items that can be preserved for longer periods can be preserved in larger batches since you have longer to consume everything. Foods that can only be preserved for short periods should be preserved in small batches so you can reasonably consume everything before it spoils. Although this may not leave you with an indefinite bounty of food, it will certainly allow you to stock more food for longer periods so that you are never without an abundance of things to eat in your pantry.

Each preservation method will have its own unique set of safety measures that should be taken into consideration when it comes to preserving food. I have included information on each of the methods written about in this book so you can feel confident in safely preserving your food every single time.

Standard Food Preservation Safety Considerations

When it comes to preserving foods, there are a few things that apply no matter what type of food preservation you are doing. Following these general standards, every time is imperative to the safety of your food. In many cases, not following these proper standards will not only reduce the safeness of your food, but it will also reduce the quality of the flavor of your food. For both of these reasons, it is imperative that you follow all of these considerations.

Keep Your Space, and Tools, Sterile

Whenever you are working with food, you should always keep your workspace sterile. When it comes to preserving food, however, you need to be even more cautious as you do not want to transfer bad bacteria into your preserves. Even a small amount of bad bacteria can rapidly multiply in a preserve recipe, leading to massive and dangerous spoilage over time.

Immediately before you begin preserving your foods, ensure that you clean everything in the hottest water possible and with proper antibacterial soaps and detergents. Keep your hands clean, as well, and prevent food from cross-contaminating other food sources. It is best to educate yourself on proper safe kitchen etiquette and follow these standards, particularly when you are preparing foods for preservation recipes.

Only Use Tools That Are In Proper Working Order

Never use any tool that is not in proper, safe working order. Most food preservation tools can be rather dangerous if they are not in proper working order, which can lead to injury, illness, or serious damage. For example, pressure canners that are not kept in proper working order can explode and damage the top of your oven, as well as the ceiling above your oven. Inspect all of your tools before use and maintain them if necessary. If you come across any tool that is broken, discard it and replace it before beginning your preservation recipe.

Be particular about tools that you use for mixing, storing, and otherwise interacting with your food, too. You might think it is only the large appliances that pose a threat, but that is not the case. A crack in a jar, a ladle that is separating from its handle, or any other cracks or breaks in small kitchen tools can harbor dangerous bacteria and contaminate your food with that bacteria. Even if you attempt to wash it properly or heat it to a safe temperature, bacteria can linger in tough cracks and become a danger to your cooking process.

Use the Freshest Foods Possible

Always use the freshest food possible, and avoid using any foods that are showing signs of spoilage or that are soon to expire. Using foods too close to their expiry date can result in them having a chance to develop harmful bacteria before you have a chance to preserve them. Once you preserve those foods, those harmful bacteria can multiply and destroy your food. In a best-case scenario, you will have an incredibly gross tasting preserve. In a worst-case scenario, you will end up with a life-threatening illness caused by eating contaminated food. Likely, however, you will end up with a lot of wasted food on your hands.

Cook Foods All the Way Through

Standard cooking rules still apply when it comes to preserving foods. With meats, in particular, you need to be confident that you have cooked them all the way through. Even if you are using less common cooking methods like smoking your meat or storing your tuna in oil, a high-quality meat thermometer should always be available when you are cooking any sort of meat or seafood product, and you should know how to use it properly. Always cook your meats all the way through to avoid having them retain harmful bacteria or parasites that can wreak havoc on anyone who may try to eat them.

Follow Exact Recipes That Match Safety Standards

Over the years, food preservation has been largely researched and perfected by professionals. Following food preservation recipes from these professionals is the best way to ensure that you are getting recipes that follow the latest safety measures for that unique type of food preservation. When you follow a recipe, always follow it exactly. Never double recipes or half recipes unless you are absolutely confident that it is safe to do so, as most recipes are made in such a way that keeps them exact and safe. Not following a recipe to its exact standards could result in you accidentally introducing harmful bacteria to your preserve, or allowing for harmful bacteria to thrive in that environment, which can lead to food spoilage and possible illness.

Properly Label Everything You Preserve

Every time you preserve something, label it. Even if you think you will remember what it is and when you preserved it, label it. Ideally, you should put a "made on" date and a "best by" date on every single thing you make. Labels ensure that you always know what is inside of a container, when it was made, and when it should be consumed by. A permanent marker and labels purchased in the canning section of any major department store should be plenty for most food preservation labeling needs. If you want to avoid cleaning stickers off of things, later on, choose labels that dissolve in water.

Use the Right Packages to Store Your Preserves

Just like you need to follow the right methods for preparing your foods, you also need to follow the correct methods for storing your foods. Avoid storing your foods in anything other than what a recipe calls for, or what you know to be absolutely safe for a certain method of preservation. Storing your food in the wrong container could expose it to excessive amounts of light, oxygen, moisture, heat, or pests. Never size up or size down on a package unless a recipe says it is safe to do so, or that particular preservation method is known for being safe to do so.

Store Your Preserves In the Right Conditions

Aside from keeping your preserves in the right containers, you also need to keep them in the right conditions. Ensure that they are kept in a proper, cool, dry place where they can be kept for long periods of time. The best space in your house to store your preserved foods is a root cellar. If you do not have a root cellar, a dark, cool closet away from the kitchen is ideal. Avoid a closet too close to your kitchen as they have a tendency to heat up from the warmth of cooking, and this can deteriorate the quality of your preserves.

Keep Track of and Consume Your Preserves In Proper Timing

After you have labeled foods and safely stored them away, you should always make a note of what you have preserved. Keeping an "inventory list" of all of the foods you have preserved is a great way for you to quickly see what you have on hand and use it before it spoils. Your inventory list should include all of the food items you have, organized by category, as well as the dates they must be consumed by. Keep those that need to be consumed quickest at the top of your list, so you know when they need to be eaten by. Plan your meals around this list and your inventory so that you use all of your preserves before they spoil. This way, you do not end up wasting any of the food you put so much effort into saving for your family.

Research a Method Before Actually Trying It

Insufficient research and improper understanding can lead to you making mistakes during the preservation process, which could result in serious illnesses or injuries. Be sure to completely read through the steps on how to complete a certain preservation method beforehand, educate yourself on the safety measures of that method, and learn the recipe before you try it. If you are extremely new to something, having the help of someone who is already experienced in the said method can be a great way for you to learn how to properly complete that method. Ensure that the person you are learning from follows the proper safety standards on that method so they can teach you those standards, too.

Safety Measures and Considerations When Canning Food

Canning is one of the most popular food preservation methods to date. With canning, the biggest risk aside from food spoilage is botulism. Improperly canned food items can carry high levels of botulism, which can infect people and cause illness and even death. Although overall death rates from foodborne botulism are low, you still need to be incredibly careful when using this preservation method to prevent the risk of infection. The lowered rate of infection is likely due to the fact that we now have a clear understanding of what safe canning looks like and how to completely eliminate botulism spores from the food contents that are stored within the can.

Always Use the Right Canning Method

Never try to can foods using a canning method other than the one described in the recipe. Attempting to water bath can recipes that should have been processed in pressure canners, or pressure can recipes that should have been processed in water bath canners, can lead to dangerous side effects. For foods improperly canned in water bath canners, the highest risk is with botulism infection. For foods improperly canned in pressure canners, the highest risk is that the pressure canner explodes and causes serious damage to your home and possibly serious injury to you or anyone around your pressure canner.

Use the Right Level of Acidity

Canning recipes call for certain levels of acidity to ensure that they are able to remain shelf-stable. The acidity of the contents in the can directly contribute to the foods' ability to remain safely preserved for lengthy periods of time. Even in low-acid foods, salts and other acidic preservatives are added to the recipe to ensure that you are storing the contents as safe as possible.

In a water bath canner, it is especially important to check your acidity levels since it is intended *only* to be used with high acidity foods. To ensure absolute safety, you should use ½ teaspoon of citric acid or two tablespoons of bottled lemon juice per quart of tomatoes. If you are canning pints, you will need one tablespoon of bottled lemon juice or ¼ teaspoon of citric acid per pint.

Use New Lids Every Time

Canning jar lids are not intended to be used more than once, unless you are using them on dry sealing methods, such as storing dehydrated foods or dry ingredients. Attempting to reuse a lid through a canning processing experience can result in it not properly sealing, which can lead to food spoilage and the introduction of botulism spores in the food you are preparing. You should always purchase new lids every time you are canning, and you should use new screw bands any time your screw bands start to rust. Otherwise, the rusted bands will be challenging to remove, and you will not be able to retrieve your contents from inside the jar. There is one long-term alternative to replacing jar seals, which is to purchase reusable jar seals. These special lids are designed to be used multiple times, though they come at a much higher price. If you are willing to foot the bill, though, they last longer and are more environmentally friendly. Be sure you use, store, and clean your reusable lids exactly to the manufacturer's recommendation to avoid accidental illness.

Always Follow the Recommended Headspace Rule

The amount of headspace recommended in a canning jar is developed to minimize the amount of air in each jar without running the risk of the jar exploding. Certain recipes, like salsa, will require larger headspace because of how much pressure they generate. Other recipes, like jam, will require less headspace because they do not produce as much pressure inside of the jar. If you leave too much headspace, the seal may not form properly, and there is too much air, which can lead to foods turning rancid inside of the jars.

Safety Measures and Considerations When Dehydrating Food

Dehydrating food is a popular method among avid food preservation fans, as well as people who love to hike. In the prepper community, this is a preferred method for storing food for as long as possible without having to follow all of the intense requirements of canning foods. As well, many simply enjoy the taste of dehydrated food.

Spread Everything Out In Shallow Layers

When you are dehydrating food, ensure to spread everything out in shallow, single layers. Piling things up or allowing the layers to become too thick can result in you having food that does not dry out fast enough, which can pose a serious threat. Food that does not dry out fast enough can start to develop harmful bacteria during the dehydrating process, which results in the food not being safe for consumption.

Store Everything In Air Tight Containers

Dehydrated food always has the ability to rehydrate by capturing moisture out of the air or anything else that may come into contact with it. Keeping your food in airtight containers is essential to avoid the food from growing dangerous bacteria. Even if you are going to be freezing your dehydrated foods, you need to keep them in airtight containers to avoid them rehydrating from the ice in the freezer. Vacuum sealed bags are the best choice for dehydrated foods that are not dehydrated to the point of being brittle.

Keep Everything In the Right Storage Conditions

Foods that are dehydrated to the point of being brittle, such as potato flakes or herbs, can safely be stored in airtight containers at room temperature. Ensure that these are kept in the smallest jar possible so that not a lot of air can get in with them or, better yet, store them in vacuum-sealed bags. If the food you have dehydrated is not brittle, such as jerky or mango slices, you will need to vacuum seal it and store it in your freezer. While these can be stored on the shelf for a period of time, room temperature will drastically reduce their lifespan, resulting in them developing harmful bacteria and turning rancid.

Rehydrated Foods Are Perishable

If you are going to be creating dehydrated foods that you plan to rehydrate later, such as potato flakes, it is important to note that the rehydrated foods are perishable. If you store your foods improperly and they become rehydrated in the process, they should be considered perishable, too. Since there is no way of knowing when the moisture was reintroduced, it would be best to toss that food out and start over.

Safety Measures and Considerations When Freezing Food

Freezing food is something most of us do, so it seems strange that there would be unique considerations for you to be mindful of when it comes to freezing foods. After all, it seems like common sense, doesn't it? There are, however, a few things you need to consider when it comes to storing food items in your freezer.

Your Freezer Must Always Be Kept At the Right Temperature

In order for your freezer to work properly, it must be kept at a temperature less than 0F. If the temperature rises about 0F, the food inside of the freezer will become dangerous to eat because it has been exposed to conditions where parasites, bacteria, and mold can begin to grow on that food. If the temperature remains below 0F the entire time, food can be stored indefinitely.

Freezing Does Not Kill Off Any Parasites, Bacteria, or Molds

Although freezing will protect your foods, it does not kill off any parasites, bacteria, or mold that have begun to grow within your food. If you freeze your food, the only thing that happens to the parasites, bacteria, and mold is that it becomes inactive. As soon as the food is defrosted, the parasites, bacteria, and mold will become active again and will continue to contaminate the meat. For certain parasites and bacteria, the cooking process will kill them off, and the food will become safe to eat. For other parasites and mold, they will continue to remain and contaminate the food regardless of what is done after they are defrosted, so the best option is to throw them away.

Food Must Be Frozen At the Right Time In Its Lifecycle

Food that is frozen at the peak of its lifecycle retains all of the nutritional content that it had before going into the freezer. It also continues to taste great, and works excellent in most dishes, as long as it has not become freezer burnt or been stored for too long. It is best to store food at the peak of its freshness, rather than at the end of its lifecycle, to ensure it remains tasty and healthy to eat.

Proper Packaging Must Happen to Prevent Food Deterioration

Properly packaging your food is essential if you are going to prevent the food from deteriorating. You need to prevent the frost and ice in your freezer from coming into direct contact with the food if you are going to be able to prevent freezer burn and discoloration. While freezer burn and discoloration do not render food inedible, and it will still be perfectly safe to eat, the tastemay be off. In some cases, extensive freezer burn and discoloration can lead to the food tasting terrible.

Safety Measures and Considerations When Brining and Salting Food

Brining is something that many people do as a way to tenderize their meat and improve the flavor. In the case of smoking food, brining or curing the meat is an essential step to higher quality meat that dehydrates and preserves better. Following proper safety measures is essential when brining food.

Always Use the Proper Salt Ratio

Always use the right amount of salt when you are brining. First and foremost, a brine will not work if there is not enough salt, as salt is responsible for the osmosis process that happens in the meat while it brines. Second, salt is what helps prevent your food from developing harmful bacteria. Foods that would generally go rotten in the fridge in just a couple days can be stored for a week or more in brine, depending on the recipe you are using. The way this is possible by using the proper amount of salt, always.

Ensure That All of Your Food Is Properly Covered

If any of your food is sticking above the salty brine, it is at risk of developing harmful bacteria or mold. In some recipes, they will suggest you salt the exposed top to prevent this from happening. In others, they may require you to use a weighted item to keep that food below the surface of the brine. Ensure that you follow these directions clearly so that you are not putting your food at risk of developing harmful bacteria.

Always Use Food-Grade Materials When Brining Food

Stainless steel, glass, and high-quality BPA-free, food-grade plastic are the best for brining. Ensure your bowls, crocks, and utensils are all made of these materials to avoid accidentally introducing harmful bacteria into your food.

Store Your Brines In the Fridge

In days gone by, brined meats were stored in cool root cellars where they stayed cold enough to remain preserved for extended periods of time. These days, the average house does not have a root cellar. As well, the root cellars are not as safe as the fridges we do have. Always store your brines in the fridge to avoid the salt brine and food from heating up to the point where it becomes dangerous to consume.

Let Your Meat Come to Room Temperature Before Cooking

Before you cook meat that has been preserved in a salt brine, always let it come up to room temperature. Cooking cold meat in cooking oil can result in serious spitting and, in some cases, explosions in your kitchen. Letting the meat warm up first not only helps tenderize it, but it also avoids you experiencing a dangerous explosion in your kitchen.

Safety Measures and Considerations When Sugaring Food

Sugaring is a fascinating method, yet it is rather hard to find information on in the modern age. Still, there are things you should know about safely sugaring food so that it stays preserved.

Always Use the Recommended Amount of Sugar

Ensure that you always use the recommended amount of sugar when sugaring foods, and only sugar foods that have specific recipes available. It might seem like you are using a lot of sugar, but that sugar content is essential for proper storage. Do not adjust the ratio. If the sugar seems like too much, use an alternative storage method.

Store Your Sugared Foods In the Fridge

Sugared foods are not shelf-stable. There is not enough acidic content in them for them to be properly preserved at room temperature. So you should never attempt to do so. The purpose of sugaring foods is to extend their available lifespan in the fridge.

Safety Measures and Considerations When Smoking Food

Smoking food is a rather popular concept in the modern age. It tastes great, and it is fairly easy to do when you know what you are doing. The following safety tips will ensure that you are able to preserve all of your foods through your smoker safely.

Use Proper Safety Equipment

Smokers get incredibly hot, and they produce a large amount of smoke. The smoke itself can become extremely hot and can produce burns if you are not careful. Always use the proper safety equipment and safety measures when you are working with a smoker. If you need to open the smoker to check on your foods, the appropriate method is to open the smoker and stand back while the excess smoke dissipates. Then, you can move forward and check your meats. Meats in a smoker do not need to be turned, so do not do this. Doing so only puts you at risk of burning yourself from unnecessarily tampering with the food.

Use the Proper Amount of Wood Chips

Using too many wood chips can lead to you having added fuel in your smoker, which can lead to a fire. Refrain from adding extra wood chips or chunks to extend your smoke time and instead commit to adding more as needed.

Do Not Leave the Smoker Unattended

Smokers are hot, and they produce flames to keep the wood smoke available for your food. For this reason, they are a fire hazard. Ensure that your smoker is on sturdy, non-flammable ground, and remain near your smoker at all times so that you can watch it and take immediate action if anything goes wrong. The user manual will tell you what to do in the event of a fire. Be sure you read that before using your smoker so that you are prepared in case of an accident.

Safety Measures and Considerations When Pickling and Fermenting Food

Pickling and fermenting foods are something you have to be careful with because you are relying on salt to prevent food spoilage. Improper methods could lead to botulism poisoning. The following tips will help you safely pickle and ferment foods.

Only Follow USDA-Approved Recipes

USDA-approved recipes provide adequate salt ratios, preparation guidance, and storage measures for pickling or fermenting foods. Ensure that you follow USDA-approved recipes as they will contain everything you need to know, including up-to-date safety measures. Avoid heritage recipes that may lack proper safety standards to avoid botulism or other types of potentially fatal food poisoning.

Use the Proper Amounts, and Types, of Salt

Like brining, pickling, and fermenting both require salt to inhibit the development of bacteria. Use the proper amounts and types of salt to avoid the development of unwanted bacteria.

Store Them Per the Recipe's Recommendation

Each pickling and fermenting recipe will have specific recommendations on how to store that recipe. Usually, it will include placing them in the fridge for several months or canning them for up to a few years. Follow the exact recommendations to ensure your food remains free of harmful bacteria after being processed.

Safety Measures and Considerations When Ash, Oil, and Honeying Food

Ash, oil, and honey are old-school methods for preserving food. They work excellently, but you need to ensure you have the right safety practices in place to avoid food spoilage, or worse.

Follow Exact Recipes

To avoid accidental illness, always follow exact recipes. Ensure your recipes are updated to include modern food safety practices so that you are not accidentally introducing bad bacteria into your foods. In addition to following exact recipes, follow exact storage methods, too. And, always use the freshest wood ash, oil, and honey available for long term storage.

Check on Your Food Regularly

Food preserved these ways should be fine, but as with any preserved food, there is always room for error. Check on your foods every so often to ensure they are still safe to consume. If they are not, remove and discard the poor quality foods, so they do not contaminate everything else.

Chapter 15

Juice and Smoothie Recipes

Being able to preserve your harvest is important. Still, it is also essential to know how to use your preserved harvest to help you create delicious meals. During times of survival, especially, eating extremely nutritional meals is important as it helps keep you sustained through stressful periods. The added nutrients will also help if you find yourself having to put more work into your daily life so that you can sustain your survival.

Although working with preserved foods is different from working with natural foods, there are still many simple and delicious recipes you can make that will allow you to enjoy the foods you have preserved.

One of the best foods you can make for yourself out of your preserved harvests is drinks. Smoothies and fresh juices are nutritious, healthy, and can give you a large amount of protein through a relatively small amount of food. They are simple to make, easy to digest, and can offer many benefits. Plus, if you are someone who tends to struggle with eating when you are stressed, smoothies and juices go down well, which makes it easier for you to load up on proper nutrition during difficult times.

Honey Berry and Spirulina Juice

Makes 1 serving of juice.

What You Need:

- 2 TBSP honey berry puree
- 1 TSP spirulina

- 1 cup of filtered water, milk, or fruit juice

How to Make It:

1. Blend all of your ingredients together in a blender. Taste it for sweetness; if you want it sweeter or more flavored, you will want to add more of your honey berry puree.

Strawberry Banana Oat Smoothie

Makes 1 smoothie.

What You Need:

- 1 frozen banana
- ½ cup frozen strawberries
- 2 TBSP soaked oats (soak overnight for softest texture)
- 1 TBSP peanut butter
- 1 cup filtered water or milk

How to Make It:

1. Cut your frozen banana into 1" slices and then place everything in your blender. Blend until you reach your desired consistency.

Frozen Peach Raspberry Dream Smoothie

Makes 1 smoothie.

What You Need:

- ¼ cup canned peaches
- ¼ cup frozen raspberries
- 1 TBSP lemon juice
- 1 cup filtered water or tropical juice
- 2 ice cubes

How to Make It:

1. Blend everything in your blender until you reach your desired consistency. If the taste of your smoothie is tart because of the raspberries, add some of the syrup out of your jar of peaches.

Superfruit Smoothie

This recipe makes 1 smoothie.

What You Need:

- ¼ cup vanilla yogurt
- ½ cup frozen berries (whatever mix you like)
- 1 cup ice
- 1 TSP coconut sugar
- Filtered water to reach your desired consistency

How to Make It:

1. Add everything to your blender. Start with just ½ cup of water and increase your water content from there until you reach your desired smoothie consistency.

Super Banana Oat Smoothie

Makes 1 smoothie.

What You Need:

- ¼ cup rolled oats
- ½ cup plain yogurt
- 1 frozen banana, sliced into 1" pieces
- ½ cup milk
- 2 TSP raw honey
- ¼ TSP cinnamon

How to Make It:

1. If you prefer softer oats in your recipes, combine the oats and milk and let them sit in the fridge overnight. Then, place everything in your blender and blend it until you reach your desired consistency. Adjust your honey content to reach your desired level of sweetness.

Ultra Green Smoothie

Makes 1 smoothie.

What You Need:

- 1 frozen banana, cut into 1" slices
- ½ cup blueberries, sliced
- ½ cup almond milk

- ½ cup baby spinach
- ¼ cup plain yogurt
- 1 TBSP nut butter, any flavor you like
- 3 mint leaves

How to Make It:

1. Add all of your ingredients to your blender and blend until you reach your desired consistency. Add extra milk to adjust thickness if you find that your smoothie is too thick.

Super Zinger Breakfast Juice

Makes 1 cup of juice.

What You Need:

- 2 carrots, peeled and chopped
- 2 beets, trimmed, peeled and chopped
- 2 apples, peeled, seeds removed and quartered
- 2 lemons, peeled, seeds removed and quartered

How to Make It:

1. Place everything in your juicer and let it run into a large glass. Add some raw honey to sweeten the drink if you are not happy with the flavor. Use your honey berry puree for an even sweeter and more distinct flavor if you wish.

Slam It Down Kale Juice

Makes 1 cup of juice.

What You Need:

- 1 stalk celery, chopped into 3” lengths
- 1 cucumber, chopped into 3” lengths
- 5 kale leaves
- 2 TBSP parsley
- 3-4 pieces of pineapple

How to Make It:

1. Cut your vegetables and fruit into sizes that will fit into your juicer. Run them through your juicer and let them fall into a large cup. Adjust pineapple for more or less sweetness in your juice, or add an apple for additional sweetness.

Chapter 16

Meal Recipes

Aside from smoothies and juices, proper meals are important to eat, too. You should be eating at least three wholesome meals every single day to ensure that you are getting plenty of sustenance to keep you going.

Because you have already done so much work in preserving your food, a lot of the time preparing a meal from your stockpile is as simple as opening a few jars and warming everything up. Of course, there are many great ways that you can pair these preserves together to ensure that you are getting a proper nutritional intake.

It is important that you focus on continuing to eat all of the important foods every single day, even if you are relying heavily on your stockpile. This means every day you should be eating hearty servings of fruits and vegetables, meat, and grains. Keeping a healthy array of proteins, fats, and carbohydrates on your plate for each meal will ensure that your body has everything you need to sustain yourself. Ensure that you regularly choose different types of vegetables, meats, and grains out of your pantry so that you get a varied diet. Variation in your diet is essential to helping you get all of the different types of vitamins and minerals that you need on a day to day basis.

One thing you should be cautious of when eating out of your preservation stockpile is portions. When you are not in a survival setting, it can be easy to overstock everyone's plates and eat more than you truly need. Aside from this being generally unhealthy, it can also lead to you depleting your stockpile much faster than you need to. Rather than letting your stockpile get depleted, make sure everyone gets proper rations. Give people the recommended daily amount, no more and no less. This way, everyone can sustain themselves, and your stockpile lasts as long as possible.

To help inspire you to get started with using your preservation stockpile in recipes, I have included a supply of recipes below. These will show you exactly how you can transform your preserved foods into delicious meals that will keep you and your family going for extended periods of time.

Mashed Potatoes and Meatloaf

Makes 2 servings of mashed potatoes and 1 whole meatloaf.

What You Need:

- 2/3 cup potato flakes
- 2/3 cup water
- ¼ cup milk
- 1 TBSP butter
- ¼ TSP salt
- 1 lb ground beef, defrosted
- ½ cup breadcrumbs
- ¼ cup tomato sauce
- 1 egg
- 2 TBSP dried herb mix

How to Make It:

1. Boil 2/3 cup of water. Add potato flakes until they are rehydrated. Remove the mixture from heat and add milk, butter, and salt. Let stand for 10 minutes, then whisk again for a fluffier texture.
2. Meanwhile, preheat the oven to 350F. In a bowl, mash together beef, breadcrumbs, tomato sauce, egg, and dried herbs. Using your hands, shape the beef mixture into a meatloaf and place it in a bread pan to bake—Bake the meatloaf for 30 minutes, or until it reaches safe internal temperatures.

Kielbasa and Sauerkraut

Makes 4 servings.

What You Need:

- 3 potatoes, peeled and diced
- 16 oz sauerkraut
- 1lb kielbasa sausage, cut into ½ inch lengths
- 1 onion, sliced thin
- ½ cup butter
- 2 garlic cloves, minced
- ½ TSP thyme
- ¼ TSP sage
- ¼ TSP black pepper

How to Make It:

1. Heat cold butter in a pan with the onions. Simmer over medium heat for 10 minutes. Add garlic, thyme, sage, and black pepper. Let it simmer for 2 minutes, so herbs become fragrant.
2. Add sauerkraut with its liquid contents, kielbasa pieces, and potato chunks. Simmer.
3. Pour mixture into a casserole dish and bake at 225F for 3 hours.

Crockpot Chili

Makes 16 servings.

What You Need:

- 1lb ground beef
- ½ onion, sliced
- 16 oz stewed tomatoes, with juice
- 8 oz tomato sauce
- 3 TBSP dried herbs
- 1 cup dried beans, any mix you like
- 3 cups soup stock, beef or vegetable

How to Make It:

1. Brown your ground beef with 1 TBSP of your dried herb mix. Add the cooked beef to your crockpot. If you'd like, you can sauté your onion in the retained beef juices. Otherwise, just add them fresh.

2. Combine all of your ingredients in a crockpot and cook over low heat until the beans are cooked all the way through. This takes about 3-5 hours.
3. Consume what you can in two days, freeze the rest in one or two serving increments.

Tuna Sandwiches

Makes 2 tuna sandwiches, plus extra bread.

What You Need:

- 1 jar of tuna in oil
- ¼ cup mayonnaise
- 3 cups flour
- 1.5 TBSP melted butter
- 1.5 TBSP sugar
- 1 TSP salt
- 1 TSP yeast
- 1 cup whole milk, lukewarm
- ¼ cup warm water, 115F

How to Make It:

1. Dissolve one-half tablespoon of the sugar in the warm water and add the yeast. Let it sit for 15 minutes until the yeast becomes frothy.
2. Mix 1.5 cups of flour, the remaining sugar, and salt together. Add the yeast mixture and stir until it for about 5 minutes, until no lumps are remaining.

3. Add the remaining flour and knead it for 10 minutes. Let it rest for 15 minutes. Transfer it into a floured bowl and let it rise for 2+ hours. The longer, the better your bread will taste.
4. Punch the dough down, knead it for 5 minutes. Place it in a greased loaf pan and let it rise for 2+ hours. Again, the longer the rise, the better the bread.
5. Bake the bread at 350F for 30 minutes, or until the top is golden brown. Remove it from the oven, rub the top with melted butter, and let it rest for half an hour.
6. Drain the tuna, flake it with your fork, and add mayo. Mix it together until you have a wet, sticky tuna salad texture.
7. Slice your bread and place tuna salad on one side of the bread. Top it with another piece and serve!
8. Store leftover bread in saran wrap and tinfoil in the cupboard and eat it within three days for the freshest flavor.

Hearty Cowboy Trails Dinner

Makes 2 meals.

What You Need:

- 1 cup dried kidney beans
- 1.5 + 2/3 cups water
- 1.5 tsp salt
- 2/3 cup potato flakes
- ¼ cup milk
- 1 TBSP butter

- ¼ TSP salt
- 4 strips beef jerky

How to Make It:

1. Soak your dried beans in 1.5 cups of water and salt overnight. Keep them in the fridge to prevent mold from developing on them.
2. When you are ready to eat, boil your beans for about 1.5 hours, or until done. Top the pot up with water as needed to prevent it from boiling off. (Alternatively, cook dried beans in 6 cups of water in the crockpot until they are done, about 3-5 hours.)
3. Thirty minutes before your beans are done, boil 2/3 cups of water. Add potato flakes and cook until they are rehydrated. Remove from heat, add milk, butter, and salt. Let stand for 10 minutes, then fluff it with a fork.
4. Serve ½ cup beans, ½ cup mashed potatoes, and two slices of beef jerky on each plate. Eat fresh.

Conclusion

Congratulations on reading *Survival 101: Food Storage.* I wrote this book to help you discover how you can safely and effectively preserve your food so that you can develop a hearty stockpile for your family. During uncertain times when grocery stores are sold out, and supply chains are running dry, knowing how to preserve your own food is important. Through preserving your own food, you ensure that your family has plenty to eat regardless of what is going on in the world around you.

For many, taking direct control over your food supply, as well as your survival, is a critical way to help you feel safe and confident during such challenging times. This way, no matter what is going on in the world, you know your family can rely on you to remain safe and healthy through the experience.

Preserving your own food can seem like a daunting task if you are new to it. I suggest you pick just two preservation methods to start with and begin there. Preserve as much as you can using these two methods. Ensure that you follow adequate safety measures and that you prepare as much as you possibly can for your family. If, by the time you are done, you feel ready to preserve more food using an alternative method, you can add that to your list. This way, you do not become overwhelmed, and you are able to confidently and safely preserve your food in minimal timing.

When you are preserving food, you must follow exact safety measures to ensure that your food remains safe to consume. Improperly preserved food can lead to serious illness, such as botulism, which can be fatal. The USDA has lists of certified recipes that are known for following proper, modern safety measures to ensure that you are using the best recipes possible. Understand that heritage recipes may have worked for people in the past, but those people did not know what we know now. As a result, many of them fell ill from consuming their food. These days, we have access to proper research, science, and technology that allows us to safely preserve and store foods without risking our health, or the health of our family. If you do choose to use a heritage recipe, compare it to USDA standards and adjust it as needed to ensure that your recipe is safe for use.

A great way to help yourself preserve recipes safely is to find an experienced person that can help you. Someone who can show you the ropes is a great way to have access to all of the support you need to properly and safely preserve your food. Ensure that the person you are learning from is up-to-date on the latest safety standards and that they use them in their own preservation methods, too, so that you can confidently consume everything you prepare. If you cannot find someone in your local area that can help you, consider finding someone online who can tell you everything you need to know. It may not be as hands-on, but it is still a great way to learn everything you can.

If you are interested in learning more about survival, I encourage you to check out my three other books on this very subject. *Survival 101: Raised Bed Gardening, Survival 101: Bushcraft,* and *Survival 101: Beginner's Guide* are all great books designed to help you take your survival into your own hands. This way, no matter what, you feel confident in your ability to keep your family safe and alive.

Before you go, I ask that you please take a moment to review *Survival 101: Food Storage* on Amazon Kindle. Your honest feedback would be greatly appreciated, as it helps me create more great content for you.

Thank you, and best of luck. Stay safe out there, friend!

Description

What happens when you head to the grocery store, and everything is sold out?

Have you ever stumbled upon the realization that your money can't buy you something that isn't for sale?

Are you wondering how you are going to feed your family?

These uncertain times have made it hard for many to gain access to their basic needs. Food, in particular, has been rapidly selling out, leaving many with scarce options when it comes to how they will feed their families. The food that remains has gone up dramatically in price and leaves many unable to reasonably fill their cupboards.

What is the solution? Food storage.

Food storage is a process that has been used for hundreds of years as a way to preserve food for extended periods. You engage in food storage already. Anytime you use your fridge or freezer or place something in your pantry, you are storing food. The key to using food storage to stockpile for emergencies successfully is to know what you need, how to preserve it, and how to use it.

The secret behind food storage is that the more processed the food is when you buy it, the more it will cost you. In many cases, the less nutritious it will be, too. Buying fresh food and preserving it yourself will save you money and allow you to feel absolutely confident that you are consuming the healthiest food possible.

Survival 101: Food Storage: A Step by Step Beginners Guide on Preserving Food and What to Stockpile While Under Quarantine is the ultimate in-depth guide that covers more than ten preservation methods, includes recipes and gives you the inside scoop on safety standards for each preservation practice. By following this book, you will discover how to stockpile food for your family in a safe and cost-effective way ***without*** the overwhelm.

Inside *Survival 101: Food Storage,* you will discover:

- 10+ preservation methods with unique recipes for food preservation
- Unique recipes you can make with your stockpile of home-preserved foods
- How to locate food to preserve, and how to save as much money as possible
- Methods for planning how much food you need to purchase and preserve for your family
- Essential USDA-certified safety measures for safe food preservation
- A step-by-step plan that walks you through exactly what needs to be done so you can preserve plenty of food for your family
- Tips on how to make preservation less overwhelming
- How to store your preserved foods for optimal freshness
- And so much more!

Even if you are brand new to food preservation, *Survival 101: Food Storage* has you covered. This in-depth guide details everything you need to know, right from the very basics, to help you safely and effectively stockpile foods for your family. By the end, you will feel like a pro. A well-fed, and well-prepared pro!

Buy your copy of *Survival 101: Food Storage: A Step by Step Beginners Guide on Preserving Food and What to Stockpile While Under Quarantine* today, and start your very own stockpile right away!

Lightning Source UK Ltd.
Milton Keynes UK
UKHW052012300921
391472UK00005B/367